# Pitman Research Notes in Mathematics Series



**Submission of proposals for consideration**
Suggestions for publication, in the form of outlines and representative samples, are invited by the Editorial Board for assessment. Intending authors should approach one of the main editors or another member of the Editorial Board, citing the relevant AMS subject classifications. Alternatively, outlines may be sent directly to the publisher's offices. Refereeing is by members of the board and other mathematical authorities in the topic concerned, throughout the world.

**Preparation of accepted manuscripts**
On acceptance of a proposal, the publisher will supply full instructions for the preparation of manuscripts in a form suitable for direct photo-lithographic reproduction. Specially printed grid sheets can be provided and a contribution is offered by the publisher towards the cost of typing. Word processor output, subject to the publisher's approval, is also acceptable.

Illustrations should be prepared by the authors, ready for direct reproduction without further improvement. The use of hand-drawn symbols should be avoided wherever possible, in order to maintain maximum clarity of the text.

The publisher will be pleased to give any guidance necessary during the preparation of a typescript, and will be happy to answer any queries.

**Important note**
In order to avoid later retyping, intending authors are strongly urged not to begin final preparation of a typescript before receiving the publisher's guidelines. In this way it is hoped to preserve the uniform appearance of the series.

**Addison Wesley Longman Ltd**
**Edinburgh Gate**
**Harlow, Essex, CM20 2JE**
**UK**
**(Telephone (0) 1279 623623)**

**Titles in this series. A full list is available from the publisher on request.**

303 Numerical analysis 1993
**D F Griffiths and G A Watson**
304 Topics in abstract differential equations
**S Zaidman**
305 Complex analysis and its applications
**C C Yang, G C Wen, K Y Li and Y M Chiang**
306 Computational methods for fluid-structure interaction
**J M Crolet and R Ohayon**
307 Random geometrically graph directed self-similar multifractals
**L Olsen**
308 Progress in theoretical and computational fluid mechanics
**G P Galdi, J Málek and J Necas**
309 Variational methods in Lorentzian geometry
**A Masiello**
310 Stochastic analysis on infinite dimensional spaces
**H Kunita and H-H Kuo**
311 Representations of Lie groups and quantum groups
**V Baldoni and M Picardello**
312 Common zeros of polynomials in several variables and higher dimensional quadrature
**Y Xu**
313 Extending modules
**N V Dung, D van Huynh, P F Smith and R Wisbauer**
314 Progress in partial differential equations: the Metz surveys 3
**M Chipot, J Saint Jean Paulin and I Shafrir**
315 Refined large deviation limit theorems
**V Vinogradov**
316 Topological vector spaces, algebras and related areas
**A Lau and I Tweddle**
317 Integral methods in science and engineering
**C Constanda**
318 A method for computing unsteady flows in porous media
**R Raghavan and E Ozkan**
319 Asymptotic theories for plates and shells
**R P Gilbert and K Hackl**
320 Nonlinear variational problems and partial differential equations
**A Marino and M K V Murthy**
321 Topics in abstract differential equations II
**S Zaidman**
322 Diffraction by wedges
**B Budaev**
323 Free boundary problems: theory and applications
**J I Diaz, M A Herrero, A Liñan and J L Vazquez**
324 Recent developments in evolution equations
**A C McBride and G F Roach**
325 Elliptic and parabolic problems: Pont-à-Mousson 1994
**C Bandle, J Bemelmans, M Chipot, J Saint Jean Paulin and I Shafrir**
326 Calculus of variations, applications and computations: Pont-à-Mousson 1994
**C Bandle, J Bemelmans, M Chipot, J Saint Jean Paulin and I Shafrir**
327 Conjugate gradient type methods for ill-posed problems
**M Hanke**
328 A survey of preconditioned iterative methods
**A M Bruaset**
329 A generalized Taylor's formula for functions of several variables and certain of its applications
**J-A Riestra**
330 Semigroups of operators and spectral theory
**S Kantorovitz**
331 Boundary-field equation methods for a class of nonlinear problems
**G N Gatica and G C Hsiao**
332 Metrizable barrelled spaces
**J C Ferrando, M López Pellicer and L M Sánchez Ruiz**
333 Real and complex singularities
**W L Marar**
334 Hyperbolic sets, shadowing and persistence for noninvertible mappings in Banach spaces
**B Lani-Wayda**
335 Nonlinear dynamics and pattern formation in the natural environment
**A Doelman and A van Harten**
336 Developments in nonstandard mathematics
**N J Cutland, V Neves, F Oliveira and J Sousa-Pinto**
337 Topological circle planes and topological quadrangles
**A E Schroth**
338 Graph dynamics
**E Prisner**
339 Localization and sheaves: a relative point of view
**P Jara, A Verschoren and C Vidal**
340 Mathematical problems in semiconductor physics
**P Marcati, P A Markowich and R Natalini**
341 Surveying a dynamical system: a study of the Gray–Scott reaction in a two-phase reactor
**K Alhumaizi and R Aris**
342 Solution sets of differential equations in abstract spaces
**R Dragoni, J W Macki, P Nistri and P Zecca**
343 Nonlinear partial differential equations
**A Benkirane and J-P Gossez**
344 Numerical analysis 1995
**D F Griffiths and G A Watson**
345 Progress in partial differential equations: the Metz surveys 4
**M Chipot and I Shafrir**
346 Rings and radicals
**B J Gardner, Liu Shaoxue and R Wiegandt**
347 Complex analysis, harmonic analysis and applications
**R Deville, J Esterle, V Petkov, A Sebbar and A Yger**
348 The theory of quantaloids
**K I Rosenthal**
349 General theory of partial differential equations and microlocal analysis
**Qi Min-you and L Rodino**
350 Progress in elliptic and parabolic partial differential equations
**A Alvino, P Buonocore, V Ferone, E Giarrusso, S Matarasso, R Toscano and G Trombetti**
351 Integral representations for spatial models of mathematical physics
**V V Kravchenko and Michael V Shapiro**

Vladislav V Kravchenko

Escuela Superior de Ingeniería Mecánica y Eléctrica del IPN, Mexico

and

Michael V Shapiro

Escuela Superior de Física y Matemáticas del IPN, Mexico

# Integral representations for spatial models of mathematical physics

CRC Press
Taylor & Francis Group
Boca Raton London New York

CRC Press is an imprint of the
Taylor & Francis Group, an **informa** business

CRC Press
Taylor & Francis Group
6000 Broken Sound Parkway NW, Suite 300
Boca Raton, FL 33487-2742



**Visit the Taylor & Francis Web site at**
**http://www.taylorandfrancis.com**

**and the CRC Press Web site at**
**http://www.crcpress.com**

**To our families:**

Kira, Kiril,

my mother Viktoria, my father Viktor,

my brother Igor, my grandmother Sofia

Faina, Marina,

my mother Genia,

and my grandson Tal

# Contents

# Chapter 0

# Introduction and some remarks on generalizations of complex analysis

**0.1** The title of the book contains two crucial terms: "spatial" and "integral representations". We live in a world which is, at least, three-dimensional, so it is quite natural to describe it mathematically via functions of all three spatial variables.

Of course many excellent books and articles treat spatial problems truly "three-dimensionally" but at the same time too many authors prefer to consider models which, being formally three-dimensional, reduce in fact to the plane situation. Then the powerful tool of complex holomorphic function theory and, especially, of integral representations, is applied.

We presume that this reduction can be explained rather by the absence of a generally accepted theory of "three-dimensional holomorphic functions" which would be adequate to the two-dimensional case.

Moreover, there actually exist several theories which are considered to be good generalizations of holomorphic function theory of one variable (= one-dimensional complex analysis). These theories are developed in different, almost non-intersecting, ways. They pose and solve quite different problems. Authors using one approach quite rarely refer to the "other area". Of course applications are also different, and even when describing the same phenomenon, they do this in different terms, so that not always is it easy to understand the common nature of these investigations.

We present in this book a theory which is a very good (in a sense which will be explained below) multidimensional analogue and generalization of one-dimensional complex analysis, and which includes surprisingly many mathematical models of important spatial physical phenomena. The heuristicity of the analogy immediately provides very useful and efficient integral representations, both for mathematical theory and for physical applications.

But let us begin with a short description of those generalizations of one-dimensional complex analysis in $\mathbb{R}^3$ and $\mathbb{R}^4$ which are the best known, and even famous.

**0.2** Let $\vec{f} \in C^1(\Omega_3; \mathbb{R}^3)$; i.e., $\vec{f}$ denotes a three-dimensional smooth vector field defined in the three-dimensional domain $\Omega_3$. If there exists a solution to the system

$$\operatorname{div}\vec{f} = 0; \quad \operatorname{rot}\vec{f} = 0 \tag{0.1}$$

in $\Omega_3$ then $\vec{f}$ is called a solenoidal and irrotational, or Laplace, or harmonic, vector field, and conditions (0.1) are considered to be a good generalization of the Cauchy–Riemann conditions in one complex variable. For one thing, it is known that every plane–parallel Laplace vector field can be identified with a holomorphic function. Both $\operatorname{div}\vec{f}$ and $\operatorname{rot}\vec{f}$ have a clear physical meaning, and thus the system (0.1) and its solutions have many important physical interpretations.

A long list of references can be given; see, for instance, the books [10, p. 81–96] by a mathematician and [136] by a geophysicist.

Now in $\Omega_3$ consider an $\mathbb{R}^4$-valued function $\vec{f} = (f_0, f_1, f_2, f_3)$ which satisfies the following system of partial differential equations:

$$\begin{cases} 0 + \dfrac{\partial f_1}{\partial x_1} + \dfrac{\partial f_2}{\partial x_2} + \dfrac{\partial f_3}{\partial x_3} = 0, \\[2ex] \dfrac{\partial f_0}{\partial x_1} + 0 - \dfrac{\partial f_2}{\partial x_3} + \dfrac{\partial f_3}{\partial x_2} = 0, \\[2ex] \dfrac{\partial f_0}{\partial x_2} + \dfrac{\partial f_1}{\partial x_3} + 0 - \dfrac{\partial f_3}{\partial x_1} = 0, \\[2ex] \dfrac{\partial f_0}{\partial x_3} - \dfrac{\partial f_1}{\partial x_2} + \dfrac{\partial f_2}{\partial x_1} + 0 = 0. \end{cases} \tag{0.2}$$

This is usually called the Moisil–Theodoresco system, having originated in the early 1930s [88]. If a solution $\tilde{f}$ to (0.2) has $f_0 \equiv 0$ then (0.2) reduces to (0.1), and hence (0.2) is also a generalization – both of one-dimensional complex analysis and of Laplace vector field theory.

The zeros in the left-hand side of (0.2) can be filled in to obtain a generalization in $\mathbb{R}^4$. The system

$$\begin{cases} \dfrac{\partial f_0}{\partial x_0} - \dfrac{\partial f_1}{\partial x_1} - \dfrac{\partial f_2}{\partial x_2} - \dfrac{\partial f_3}{\partial x_3} = 0, \\[2ex] \dfrac{\partial f_0}{\partial x_1} + \dfrac{\partial f_1}{\partial x_0} - \dfrac{\partial f_2}{\partial x_3} + \dfrac{\partial f_3}{\partial x_2} = 0, \\[2ex] \dfrac{\partial f_0}{\partial x_2} + \dfrac{\partial f_1}{\partial x_3} + \dfrac{\partial f_2}{\partial x_0} - \dfrac{\partial f_3}{\partial x_1} = 0, \\[2ex] \dfrac{\partial f_0}{\partial x_3} - \dfrac{\partial f_1}{\partial x_2} + \dfrac{\partial f_2}{\partial x_1} + \dfrac{\partial f_3}{\partial x_0} = 0 \end{cases} \tag{0.3}$$

is sometimes connected with the name of R. Füter (see [37]) for reasons which are explained below. In fact, (0.3) has appeared in the works of Gr. Moisil and N. Theodoresco also.

However, the best–known generalization of complex function theory (which preserves the complex structure, by the way: complex numbers are not even mentioned above) is $n$-dimensional complex analysis. We will take $n = 2$ only. Let $x = (x_1, y_1, x_2, y_2) \in \mathbb{R}^4$, $z = (z_1, z_2) \in \mathbb{C}^2$, and we identify $x$ with $z : z_1 = x_1 + iy_1, z_2 = x_2 + iy_2$. Let $f \in C^1(\Omega_4, \mathbb{C})$; $f$ is called holomorphic in $\Omega_4 \subset \mathbb{R}^4 \approx \mathbb{C}^2$ if

$$\begin{aligned} \partial_{\bar{z}_1} f = \frac{\partial f}{\partial \bar{z}_1} &:= \frac{1}{2}(\frac{\partial}{\partial x_1} + i\frac{\partial}{\partial y_1})f = 0, \\ \partial_{\bar{z}_2} f = \frac{\partial f}{\partial \bar{z}_2} &:= \frac{1}{2}(\frac{\partial}{\partial x_2} + i\frac{\partial}{\partial y_2})f = 0 \end{aligned} \tag{0.4}$$

in $\Omega_4$.

Thus these four theories all generalize one-dimensional complex analysis to three and four dimensions, at least in the sense that all their definitions include the definition of a holomorphic function as a special case.

In spite of this, all four generalizations greatly differ from one-dimensional complex analysis. Consider, for instance, Laplace vector field theory: it does not have such far-reaching developments, and many facts in it are only slightly similar to those in one-dimensional complex analysis.

In our opinion, the main cause of the mathematical difficulties arising here is hidden inside the formal shortcomings of the traditional operations over vectors, that is, the scalar and vector products: for instance the second one is not associative and has no unit. Moreover, it is known that in $\mathbb{R}^3$ it is impossible in principle to introduce a multiplicative structure with good enough properties that we could call elements of $\mathbb{R}^3$ "numbers". That is why the rules of the scalar and vector multiplications are so complicated, and that is why for $\mathbb{R}^3$-valued functions we lose almost all algebraic properties of $\mathbb{C}$-valued functions.

When dealing with the systems (0.2) and (0.3) we should take into account that, in fact, we are considering $\mathbb{R}^4$-valued functions but operating with them with the aid of $4 \times 4$ matrices; i.e., again algebraically the situation is not very similar to that of $\mathbb{C}$-valued functions.

As for the system (0.4), it is well known that the structure, the main features of the corresponding theory, are paradoxically different from those of one–dimensional complex analysis.

**0.3** One of the main goals of this book is to present a theory which is more adequate the multidimensional situation, at least in the sense that it includes all the theories described above (and several more theories) but which avoids almost completely their disadvantages. This requires formulating some criterion for comparing the theories.

To do this let us analyse the basic elements which underlie one-dimensional complex analysis.

We stress in particular one algebraic and two analytic facts. By algebraic fact we mean, of course, the excellent algebraic structure of $\mathbb{C}$, the range of our functions. In particular, complex conjugation provides the invertibility of any non-zero complex number. (By the way, the commutativity of the multiplication is useful and pleasant to work with, but not of great importance.) In $\mathbb{R}^3$ any possible multiplication is too far from the complex one by its algebraic properties. In $\mathbb{R}^4$ the quaternionic multiplication loses only the property of commutativity.

The algebraic fact provides one of the analytical facts. Let us use the standard notation

$$\frac{\partial}{\partial z} =: \partial := \frac{1}{2}(\frac{\partial}{\partial x} - i\frac{\partial}{\partial y}), \quad \frac{\partial}{\partial \bar{z}} =: \bar{\partial} := \frac{1}{2}(\frac{\partial}{\partial x} + i\frac{\partial}{\partial y}); \tag{0.5}$$

then properties of complex numbers alone give us the factorization of the two-dimensional Laplace operator $\Delta_2$:

$$\partial \cdot \bar{\partial} = \bar{\partial} \cdot \partial = \frac{1}{4}\Delta_2. \tag{0.6}$$

If $\bar{\partial} f = 0$ in $\Omega_2$ then $f$ is called holomorphic. If $\partial f = 0$ in $\Omega_2$ then $f$ is called antiholomorphic.

Let us consider now the following problem: for which sets $\{a, b, c, d\} \subset \mathbb{C}$ does the factorization

$$\partial \cdot \bar{\partial} := (a \cdot \partial_x + b \cdot \partial_y) \cdot (c \cdot \partial_x + d \cdot \partial_y) = \Delta_2$$

hold? The general solution is given by

$$a = \pm\frac{1}{di}; \quad b = \frac{1}{d}; \quad c = \pm di, \quad d \in \mathbb{C}\backslash\{0\}.$$

Hence we can take

$$\partial \; := \; \pm\frac{1}{di}(\partial_x \pm i\partial_y),$$

$$\bar{\partial} \; := \; \pm di(\partial_x \mp i\partial_y),$$

for any $d \neq 0$. Thus for complex functions there exist exactly two essentially different types of "holomorphy".

Now let us return to the analytic facts lying at the basis of holomorphic function theory.

The second analytic fact is, in some sense, absolute. This is Green's (or the two-dimensional Stokes) formula, which has a natural analogue in any dimension: for appropriately smooth $f$ and $\Omega_2$ with boundary $\Gamma = \partial\Omega_2$,

$$\int_\Gamma f d\tau = \int_{\Omega_2} \bar{\partial} f d\bar{\tau} \wedge d\tau = 2i \int_{\Omega_2} \bar{\partial} f d\zeta d\eta. \tag{0.7}$$

From (0.6) and (0.7) the main theorems follow immediately. Indeed, let $\theta_2$ denote the fundamental solution of $\Delta_2$:

$$\theta_2(z) := \frac{-1}{2\pi}\ell n|z|^{-1}, \; z \neq 0; \tag{0.8}$$

then the $\partial$-derivative of $\theta_2$ gives the fundamental solution of the Cauchy–Riemann operator $\bar{\partial}$, or the Cauchy kernel of one-dimensional complex analysis:

$$\mathcal{K}_{\mathbb{C}}(z) := \frac{2}{i}\partial[\theta_2](z) = \frac{1}{2\pi i} \cdot \frac{1}{z}, \; z \neq 0. \tag{0.9}$$

Assuming in (0.7) $f = f_1 f_2$ we have

$$\int_{\Gamma} f_1 f_2 d\tau = 2i \int_{\Omega_2} (\frac{\partial f_1}{\partial \bar{\tau}} \cdot f_2 + f_1 \cdot \frac{\partial f_2}{\partial \bar{\tau}}) d\zeta d\eta. \tag{0.10}$$

Making use of some standard tricks by cutting out a neighbourhood of the singularity of $\mathcal{K}_{\mathbb{C}}(\tau - z)$, we can replace $f_1(\tau)$ (or $f_2$) by $\mathcal{K}_{\mathbb{C}}(\tau - z)$ which gives for $\forall z \in \Omega_2$

$$f(z) = \int_{\Gamma} \mathcal{K}_{\mathbb{C}}(\tau - z) f(\tau) d\tau - 2i \int_{\Omega_2} \bar{\partial} f \cdot \mathcal{K}_{\mathbb{C}}(\tau - z) ds_{\tau}. \tag{0.11}$$

This is the Borel–Pompeiu (= Cauchy–Green) formula, and it holds for smooth functions. It should be stressed that the proof is based on the fact that the Cauchy kernel $\mathcal{K}_{\mathbb{C}}$ is a holomorphic function (away from the origin). This is a direct corollary of (0.6), (0.8) and (0.9).

Now if $f$ is holomorphic in $\Omega_2$, then (0.7) reduces to the Cauchy integral theorem, and (0.11) reduces to the Cauchy integral formula.

It is very important to take into account the following fundamental properties of the Cauchy kernel:

- it is holomorphic,

- it is reproducing; that is, the Cauchy integral formula holds,

- it is universal; that is, the Cauchy representation works in an arbitrary domain with non-onerous restrictions on the shape of its boundary.

There exist no such kernels in several complex variables, for example.

The calculations given above have a somewhat algebraic character: if $\bar{\partial}$ denotes, instead of the operator from (0.5), some other differential operator in $\mathbb{R}^m$ with the analogues of properties (0.6) and (0.7), then we can simply try to follow the same line of reasoning to obtain analogues of (0.9), (0.11), etc.

Thus if we desire to get a "close" generalization of one–dimensional complex analysis then we should bear in mind the following principles:

- the functions under consideration should have ranges in some appropriate algebra,
- the differential operators which generalize the operators $\partial$ and $\bar{\partial}$ should be constructed with the help of elements of this algebra, and they should factorize the corresponding Laplace operators.

We hinted above that the four theories described do not satisfy these requirements: each of them loses something. We do not discuss here how important, useful, or interesting they are, but only how they agree with these two principles. Of course one can attempt to generalize some other definition of holomorphic function, but all these definitions are equivalent, and all of them generate on some step the operators $\partial$ and $\bar{\partial}$. If some generalization does not give analogues of $\partial$ and $\bar{\partial}$, or if these analogues do not give a factorization of the Laplace operator, then almost certainly we will lose too many properties of holomorphic functions.

0.4 A theory containing the four described above has existed for at least sixty years, and indeed it satisfies both principles. We describe it now.

Let $\Omega$ denote a domain in $\mathbb{R}^3$ or $\mathbb{R}^4$, and let $\mathbb{H}$ be the skew–field of real quaternions. The reader can find its principal properties in Section 1, and a more detailed survey in Appendices A1–A3.

Introduce a differential operator (sometimes called the Moisil–Theodoresco operator)

$$D_3 := \sum_{k=1}^{3} i_k \frac{\partial}{\partial x_k} \tag{0.12}$$

defined on $C^1(\Omega_3, \mathbb{H})$ (here $i_1, i_2, i_3$ are the quaternionic imaginary units). Then the action of $D_3$ can be rewritten in vectorial terms: if $f = f_0 + \vec{f} \in C^1(\Omega_3, \mathbb{H}) = C^1(\Omega_3; \mathbb{R} \oplus \mathbb{R}^3)$ then

$$D_3[f] = \operatorname{grad} f_0 - \operatorname{div} \vec{f} + \operatorname{rot} \vec{f}. \tag{0.13}$$

We shall call quaternion-valued functions from $\ker D_3$ "hyperholomorphic" or quaternionic holomorphic (synonyms: regular, monogenic, spatial holomorphic vectors). Conditions (0.1) arise when we limit (0.13) to hyperholomorphic functions with vanishing (or constant) scalar part $f_0$.

Conditions (0.2) arise if we rewrite the equation $D_3[f] = 0$ coordinate-wise, i.e., if we forget for the moment the multiplicative structure in $\mathbb{H}$.

For functions of four variables, hyperholomorphic functions can be defined by the operators $D_4 := \sum_{k=0}^{3} i_k \frac{\partial}{\partial x_k}$ or

$$\widetilde{D}_4 := \frac{\partial f}{\partial x_0} + i_1 \frac{\partial f}{\partial x_1} + i_2 \frac{\partial f}{\partial x_3} - i_3 \frac{\partial f}{\partial x_3} = 2(\frac{\partial f}{\partial \bar{z}_1} + i_2 \frac{\partial f}{\partial \bar{z}_2}).$$

Then the system (0.3) is nothing more than the matrix representation of the equation $D_4[f] = 0$, and the usual holomorphic functions of two complex variables (i.e., solutions of (0.4)) form a proper subset of the set of holomorphic functions generated by $\tilde{D}_4$. All this means that the four quite different generalizations of the notion of holomorphic function of one variable are, in fact, different representations of some quaternionic notion which covers all of them.

So far this only means that we can consider one theory instead of four, but if the new theory does not bring sufficient advantages it would not be worth being developed. The theory of quaternionic holomorphic functions, or quaternionic analysis, inherits the structural analogy from one-dimensional complex analysis. This is a consequence of the following two facts. First, the algebraic structure of $\mathbb{H}$ is quite similar to that of $\mathbb{C}$, in particular, the quaternionic conjugation $\bar{a} := a_0 - \sum_{k=1}^{3} a_k i_k$ allows us to factorize the quadratic form $|a|^2 := \sum_{k=0}^{3} a_k^2$ by $|a|^2 = a \cdot \bar{a} = \bar{a} \cdot a$. Second, as in $\mathbb{C}$, this algebraic fact provides a factorization of the corresponding Laplace operator: if $D$ denotes any of $D_3, D_4, \widetilde{D}_4$ and if $\bar{D}$ denotes its quaternionic conjugate, then for $m \in \{3, 4\}$

$$D \cdot \bar{D} = \bar{D} \cdot D = \Delta_{\mathbb{R}^m}. \qquad (0.14)$$

Thus the possibilities for exploiting successfully the scheme given in Subsection 0.3 are quite high; in fact it has been carried out several times: from the pioneering works of G. Moisil and N. Theodoresco [87, 88] and of R. Füter and his school [37], to the recent book by K. Gürlebeck and W. Sprössig [46] and a series of articles of M. Shapiro and N. Vasilevski [126, 127, 111, 112, 113, 114].

The same idea serves perfectly in higher dimensions also when the skew–field of quaternions is replaced by an appropriate Clifford algebra. We refer the reader to the books [18, 29, 40, 101, 85] and to the references therein.

**0.5** The above discussion was concerned with harmonic generalizations of one–dimensional complex analysis: quaternion–valued functions generate partial differential operators with quaternionic coefficients which preserve the factorization of the Laplace operator.

It required significant effort to comprehend that, perhaps, it is possible to take something other than the Laplacian and nevertheless preserve the main line of reasoning.

The Helmholtz operator $\Delta_\lambda := \Delta + \lambda$, $\lambda$ a constant, seems to be the nearest such "something". To study functions from the $\ker \Delta_\lambda$ (they are sometimes called metaharmonic) is the same as studying eigenfunctions of the Laplace operator, which emphasizes the "naturality" of such generalization.

Let $D_3$ denote the Moisil–Theodoresco operator. Then $\bar{D}_3 = -D_3$ and hence (see (0.14))

$$D_3 \cdot \bar{D}_3 = -D_3^2 = \Delta_{\mathbb{R}^3}, \qquad (0.15)$$

that is, $D_3$ is a square root of $-\Delta_{\mathbb{R}^3}$. Observe that by definition (0.12), $D_3$ has no scalar part; i.e., algebraically it acts as a "purely vectorial quaternion". All quaternionic square roots of a negative number are purely vectorial quaternions, and one may consider $-\Delta_{\mathbb{R}^3}$ to be analogous to a negative number. Continuing this analogy, let $\alpha$ be a square root of $\lambda : \alpha^2 = \lambda$. Then we have

$$(\alpha + D_3)(\alpha - D_3) = \lambda + \Delta_{\mathbb{R}^3}, \tag{0.16}$$

that is, we have factorized the three-dimensional Helmholtz operator, and so we can try to develop a function theory corresponding to the operators $(\alpha + D_3)$ and $(\alpha - D_3)$, analogues of $\bar{\partial}$ and $\partial$ respectively.

The first work in this direction belongs, as far as we know, to E. Obolashvili [94] who treated the case $\lambda$ real and negative, $\alpha$ a purely imaginary quaternion. Much later, Huang Liede [77] obtained the same (in principle) results. Both of them used a technique of the matrices related to the regular representations of the quaternions (like systems (0.2), (0.3) above), not the quaternions themselves.

The study of the case $\lambda > 0$, $\alpha > 0$ was initiated in the work by K. Gürlebeck [45], and examined in detail in the book of K. Gürlebeck and W. Sprössig [46].

Some special problems of both theories have been treated in the works of the authors [105, 106, 107, 64].

Although both theories ( $\lambda > 0$ and $\lambda < 0$ ) develop in quite similar ways, formally they are completely different and have no intersection. The desire to construct a theory combining them and including the case $\lambda \in \mathbb{C}$ also important for applications led us to consider the Helmholtz operator and its factorization for $\lambda$ a complex quaternion; $\lambda = \sum_{k=0}^{3} \lambda_k i_k, \quad \lambda_k \in \mathbb{C}$.

We have expressed some purely mathematical reasons which led us to a series of works [71, 69, 70] and to this book, but there have been some "applied" reasons also. We knew many very interesting applications of harmonic quaternionic analysis to mathematical physics. In [51, 52] there was demonstrated the possibility of reformulating classical electrodynamics in terms of quaternions. In [42, 4, 75] some spatial problems of elasticity theory were solved by means of quaternionic analysis methods. The relationship of a number of hydrodynamical and geophysical models with quaternionic integral and differential equations was shown in [125, 137, 136]. In [9, 35] (see also references there) the symmetrical analysis of the Dirac equation for a free particle with spin 1/2 and non-zero rest mass was essentially simplified after using its quaternionic reformulation. The analytical biquaternionic approach to the Dirac equation for a massless spinor field was proposed in [119]. We would like to mention two remarkable books [8, 46] which appeared almost simultaneously and show different interesting aspects of the applications mentioned.

The expectation that our metaharmonic quaternionic analysis would give us more applications was fully confirmed: we have found, for instance, that both time-harmonic electromagnetic and time-harmonic spinor fields are excellently described by $\alpha$-holomorphic functions. It is paradoxical here that spinor fields require, for their description, quaternionic (not complex) values of the parameter $\alpha$.

Moreover, $\alpha$-holomorphic function theory has proved to be applicable to a very rich class of physical models which are unified by the idea of hypercomplex factorization (Chapter 4).

**0.6** Besides this introduction, the book contains sixteen sections arranged in four chapters, and four appendices.

In the first chapter we develop the quaternionic holomorphic function theory corresponding to the Helmholtz equation. This requires, first of all, describing basic properties of quaternions, real and complex, which is done in Section 1.

In Section 2 we introduce the notion of the Helmholtz operator with a complex–quaternionic wave number. In particular, when the latter is a complex number, the operator coincides with the "usual" Helmholtz operator. The central results of Section 2 are Theorem 2.13 and Theorem 2.18 on a decomposition of the null set of the Helmholtz operator with an arbitrary complex-quaternionic parameter (including zero). We represent the null set of the Helmholtz operator as a direct sum of null-sets of quaternionic differential operators ${}^{\psi}\mathcal{D}_\alpha$ which generalize the one-dimensional Cauchy–Riemann operator. Thus we get, in fact, a family of such representations depending on

$\psi$ and $\alpha$, $\psi$ being an orthonormal basis in $\mathbb{R}^3$ and $\alpha$ being a (complex-quaternionic) square root of the wave number $\lambda$. Let us stress that in one–dimensional complex analysis there is no exact analogue of these decomposition theorems.

For $\alpha$ purely vectorial, the decomposition theorems were obtained in [107], the general case $\lambda \neq 0$ was published in [71], the case $\lambda = 0$ in [66]. One more interesting thing here is the fundamental solution to the Helmholtz operator with a general, complex–quaternionic, parameter.

The main object of the book, the notion of $\alpha$-holomorphic function, is introduced in Section 3. We discuss there some basic properties of such functions as well as how to construct in explicit form the fundamental solution of the operator ${}^{\psi}\mathcal{D}_{\alpha}$. In spite of the existence of some general formulas for fundamental solutions of rather general differential operators with constant coefficients (e.g., [57]), these contain some convolution operators or series, and thus they are not too pleasant to work with. Thus it is very important to have a "living" fundamental solution, the most explicit form of it, for specific operators. The fundamental solution of $D_{\alpha}$ for an arbitrary complex quaternionic $\alpha$ was calculated in [71, 70]. The content of Subsections 3.5–3.9 was taken from [127, 114]. In Subsection 3.24 we discuss the notion of Füter–type operators, in particular, of the Füter operator the essential properties of which are taken from [114], and we refer the reader to [82, 83, 112, 114] for more details and numerous applications.

In Section 4 the results of previous sections are used to obtain the main integral formulas and theorems for $\alpha$-holomorphic functions. Here the central fact, of course, is Theorem 4.17 containing the generalizations of the Borel–Pompeiu formula, Cauchy's integral formula, Morera's theorem and a theorem on a right inverse operator corresponding to ${}^{\psi}\mathcal{D}_{\alpha}$ with an arbitrary complex quaternionic $\alpha$. Let us mention also Theorem 4.20 and Theorem 4.21 giving new integral representations for harmonic and metaharmonic functions using the decomposition theorems from Section 2, and Theorem 4.17.

Boundary properties of $\alpha$-holomorphic functions are studied in Section 5. Theorem 5.3 shows that for an arbitrary $\alpha \in \mathbb{H}(\mathbb{C})$ there exists a generalization of Plemelj–Sokhotski's formulas for null–solutions of ${}^{\psi}\mathcal{D}_\alpha$. This allows us to prove the involutiveness of the corresponding singular integration operator ${}^{\psi}S_\alpha$ (Theorem 5.5) as well as to obtain the sufficient and necessary conditions for the existence of an $\alpha$-holomorphic extension of a given function on a surface (Theorem 5.7). Sections 5.11–5.31 are devoted to the analogues of the Hilbert operator and Hilbert formulas for $\alpha$-holomorphic functions. Paradoxically, the situation of $\alpha = \alpha_0 \in \mathbb{C}$ is much more complicated than that of $\alpha = \vec{\alpha}$. It is enough to compare Definition 5.18 and Definition 5.27 of the corresponding Hilbert operators. Nevertheless the Hilbert formulas are obtained in both cases (Theorem 5.19, Theorem 5.28) as well as some corollaries of them. The fine point here is that the case of $\alpha = \vec{\alpha}$ requires the usual functional spaces $C^{0,\epsilon}$ and $L_p$ while in the case of $\alpha = \alpha_0 \in \mathbb{C}$ we are forced to work in the Lizorkin space. Part of the results of Sections 4 and 5 related to an arbitrary complex-quaternionic $\alpha$ were published in [69, 70]. The Hilbert formulas and the Hilbert operator for such $\alpha$ are completely new.

In Section 6 we examine some boundary value problems for $\alpha$-holomorphic functions, such as the Dirichlet problem, reconstruction of an $\alpha$-holomorphic function in a half–space by its two given components on a hyperplane, an analogue of the Riemann problem. Here we use the results from [63, 113, 114, 68].

Since each quaternion is formally a sum of a scalar and of a vector, and the quaternionic product contains both scalar and vector products of two vectors, in Section 7 we give a vectorial reformulation of some complex–quaternionic objects introduced, as well as of some results obtained in the previous sections. For example, we discuss here the connection between the integral representation for metaharmonic functions from Section 4 with the known criterion of a vector–valued function belonging to the null set of the Helmholtz operator.

As was mentioned before, $\alpha$-holomorphic theory has found numerous and deep relations with various areas of physics: electrodynamics, spinor field theory, etc. Moreover, quaternionic holomorphic functions in the sense of R. Füter have proved to be very useful for studying non–linear objects also, for instance, the self-duality equation of gauge theory.

The rest of the book (Chapters 2, 3, 4) deals with these relations.

Chapter 2 deals with some applications to electrodynamics. Since J. C. Maxwell wrote and published his famous equations, they have been investigated in a large number of works. There probably exist no fewer works generalizing these equations in many diverse directions. There is no need to spend time explaining the reasons for such phenomena, they are evident: the importance of the subject.

At the same time, the necessity of studying the equations for more than a century bears witness to the absence of a sufficiently complete theory for Maxwell's equations.

Various hypercomplex approaches to studying the classical Maxwell equations have more than a century of history, starting from the work of Maxwell himself (which sometimes surprises both mathematicians and physicists).

There exists a reformulation of these equations in vacuum in quaternionic terms (see e.g., [116, 22, 51, 97, 52, 8, 58]) which allows some fundamental physical laws to be rewritten in a space-saving form. This is the very case in which such a phenomenological simplification is a real discovery influencing the development of a physical theory.

Formally, this leads to a partial differential operator with quaternionic coefficients which has a null set containing all solutions to the Maxwell equations. So the problem arises as to whether this null set possesses a well-developed function theory, in the sense described above.

For many specific radio engineering, hydroacoustical, geophysical models it is natural and quite sufficient to limit the study to the time-harmonic case (see e.g., [48, 92, 28, 32, 50, 54] and many other books and articles). The main reason is contained, in fact, in the Fourier analysis together with the principle of superposition: an electromagnetic wave is a superposition (or, in other words, a linear combination, finite or denumerable) of elementary, periodic-in-time waves. Of course, technically the time-harmonic case is simpler but at the same time, even for that case many profound physical properties are not understood and explained. For example: the behaviour of the electromagnetic vector field near and on the boundary of a spatial domain until now has had a far from sufficiently complete description. Most of what is known is contained in [27], see also "less rigorously mathematical" [136] where many interesting results and ideas can be found. It is written in traditional vectorial language (as is [27]) but there are some important hints as to how to develop the corresponding hypercomplex approach.

The main difference between our work and those mentioned above and the others in this direction, consists of the following. We not only rewrite the Maxwell equations in a space-saving form (which generally speaking would not give essentially new information) but also with the aid of a simple matrix transform we imbed time-harmonic electromagnetic field theory into the theory of $\alpha$-holomorphic functions described in Chapter 1.

After a brief discussion in Section 8 of the classical Maxwell equations and electromagnetic potentials we give in Section 9 the main relation between electromagnetic fields and $\alpha$-holomorphic functions: both electric and magnetic components of a time–harmonic electromagnetic field are linear combinations with complex coefficients of two functions, one of which is $\alpha$–holomorphic and the other is $(-\alpha)$–holomorphic. All results of Section 10 and Section 11 are based on this relation. Among others we distinguish Theorem 10.6 giving the useful representation for an impulse of an electromagnetic field via an integral over the boundary (obtained in [71]), Theorem 11.3 giving a criterion for the existence of an electromagnetic extension for a pair of $\mathbb{C}^3$-valued functions given on the boundary (obtained in [60]), as well as the analogue of the Hilbert formulas for electromagnetic fields (Theorem 11.8) published here for the first time.

Chapter 3 begins with the biquaternionic reformulation of the classical Dirac equation describing a free particle of spin 1/2 and non-zero rest mass. We establish an intimate, quite simple, connection between the Dirac operator and the operator $D_\alpha$. Let us emphasize that here $\alpha$ must be a complex quaternion. In Section 12 we used the results of [63, 64, 65, 67].

The connection established in Section 12 allows us to obtain in Section 13 the analogues of some theorems from Chapter 1 for massive spinor fields. As in Chapter 2 here also we do not aim at transferring all results of Chapter 1 to the solutions of the Dirac equation, but to show some interesting (in our opinion) examples of how it can be done.

Our technique works well for domains with boundary, and allows us to solve a number of boundary value problems (see the examples in Section 6). At the same time the boundary value problems for the Dirac equation appear more and more often in the literature in connection with Casimir's effect [90] and the so-called bag models describing the phenomenon of quark confinement (see, e.g., [26, 124]). There is no complete solvability theory for such problems, but there are some special solutions in some special cases. In Section 14 we show some first applications of our theory to the investigation of bag models. We reduce the MIT bag model to a boundary singular integral equation that can be interpreted as a solvability criterion for the MIT bag model. Some of the results of Section 14 were published in [64].

Section 15 (from Chapter 4) is devoted to the notion of hypercomplex factorization (Definition 15.2). We show that many operators of mathematical physics such as the Laplacian, the Helmholtz operator, the Lamé operator, the operator of steady oscillation, the operator generating the equations of the statics in the theory of moment elasticity, etc., admit a hypercomplex factorization. For such operators we give a general scheme (Theorem 15.9) for reducing the corresponding Dirichlet problem to a pair of boundary value problems for $\alpha$-holomorphic functions, thus diminishing the order of the derivative. Theorem 15.9 is in fact the generalization of results of [45, 46, 4] where there appeared the ideas on the hypercomplex factorization of the

Dirichlet problem for the Laplace, Helmholtz and Lamé equations. It gives a general perspective on these and other results and was obtained in [61]. Then we show that the notion of hypercomplex factorization allows one to generalize the results of Section 2 on the decomposition of the null set of operators (Theorem 15.16 is published here for the first time).

In the last section, Section 16, we use the "procedure of factorization" from the theory of symmetry groups (see the description in Subsections 16.4, 16.5) for the Füter–type equation. Let us emphasize that this procedure has nothing in common with a decomposition of something into the product of several factors and it is a pity that we have to use the overloaded term "factorization" again but with different meaning. We follow here commonly accepted terminology (see, e.g., [128, 122]). With the help of this procedure we obtain the solutions to some systems of non-linear equations. It is worth noticing that among them is the self–duality equation, so we obtain a new class of instantons. Section 16 is based on the results of [58].

In Appendices A1–A3 we give an ample collection of algebraic properties of the set of quaternions, real and complex. In contrast with the main text of the book, all the material in Appendices A1–A3 is well known but is not easy to find in a single source. So we believe it will be convenient for the reader to see them described in detail. We hope also that the more profound acquaintance with algebraic properties of quaternions will necessarily result in new analytic facts; i.e., in new advances in quaternionic holomorphic function theory.

In Appendix A4 we give a sketch of the theory of $\alpha$-holomorphic functions of two (not three) real variables. Being, obviously, a particular case of the theory constructed in this book, it has some interesting specific features which are described here.

**0.7** We can say that only the first chapter truly belongs to "pure mathematics". All the rest is, in a definite sense, a direct application of hyperholomorphic function theory. Of course, we do not say that "direct" means "trivial". We still find it striking that hyperholomorphic function theory has proved to be a mathematical tool which allows us to treat so many completely different objects from mathematical and theoretical physics: electromagnetic fields, spinor fields, the self-duality equation and others (not to mention, for instance, harmonic vector fields and the Moisil–Theodoresco system, the relation of which to quaternionic holomorphic function theory is well known). This is, in our opinion, the main point of this book: it offers, develops, and applies a quite new mathematical theory for treatment of an ample series of spatial problems of electrodynamics, particle physics, and elasticity theory. This mathematical theory proves to be as powerful for solving spatial problems of mathematical physics as complex analysis is for solving planar ones. As a matter of fact we provide a unified mathematical approach to a wide spectrum of important physical problems. Both the mathematical tool and the manner of using it were elaborated essentially by the authors, and the major part of the results of the book are ours.

We believe that the book [46] is the closest to our work. It was the first book which actively and successfully used the hypercomplex analysis methods for solving boundary value problems. It has incorporated many essential results which, before, had been represented by journal articles only. In the 1990s this line of hypercomplex function theory as well as its numerous applications has been developing intensely, and we hope that our book reflects important achievements of the period in this direction. There are several books [18, 29, 40, 85] devoted to "pure" hypercomplex analysis. In our opinion, they minimally intersect our book. Moreover, the interrelation should be called "mutually complementary": our work shows explicitly how and where pure hypercomplex analysis can be used in applied areas of science.

The breadth of topics covered as well as the novelty of the results has caused us to limit ourselves instead of developing both the mathematical theory and its consequences for all the applications described above, we decided to present only the main ideas and main results. One of our purposes is to popularize the effectiveness of quaternionic analysis, as many mathematicians–analysts are still quite skeptical about it. Here are some examples.

One of the great mathematicians of this century, L. Pontriagin, in the book [98, p. 4] wrote: "Because of the absence of the commutativity of multiplication it appeared to be impossible to construct a function theory of a quaternionic variable", and then again on page 97: "... the absence of the commutativity does not give an opportunity to develop a theory of quaternionic functions."

In [76, p. 213], the well-known expert in multidimensional complex analysis A. Kytmanov quotes a theorem on properties of the singular integral with the Bochner–Martinelli kernel, i.e., a purely "complex analysis" fact, which had been obtained by M. Shapiro and N. Vasilevski by quaternionic analysis methods. Though citing the source of the result A. Kytmanov does not mention the quaternionic tools by means of which it was obtained, and gives no "complex function theory" reasons for this property, whereas the quaternionic analysis explanations are clear, transparent and heuristic.

The authors will consider this book a success if a number of its readers decide to use the hypercomplex analysis techniques presented here in the investigation of physical phenomena.

**0.8 Acknowledgements** We are grateful to the *Consejo Nacional de Ciencia y Tecnologia (CONACYT, Mexico)* which supported us through project 1821-E9211 as well as to *the International Association for Cooperation with Scientists from the former Soviet Union* which supported the first-named author through project No. INTAS–93–0322.

We would like to express our gratitude to a number of people for their influence on the making of this book. The stimulating and inspiring discussions with Prof. Viktor Kravchenko on the subject of the book go back to, at least, 1989. The authors' research interests and tastes had been shaped in the fruitful and creative atmosphere of Prof. G. Litvinchuk's Seminar "Boundary value problems and integral equations" (Odessa, Ukraine) and of the Seminar "Linear operators" created later at the Odessa State University by N. Vasilevski and M. Shapiro. The comments, criticism, and advice of the Seminars' participants were always extremely valuable for us. The second-named author would like to mention especially a collaboration with Prof. N. Vasilevski over many years. The books [18] by F. Brackx, R. Delanghe, and F. Sommen and [29] by R. Delanghe, F. Sommen, and V. Soucek, as well as the book [46] by K. Gürlebeck and W. Sprößig have demonstrated the beauty and power of hypercomplex analysis. The reading of them and numerous discussions with their authors stimulated our investigations. Prof. M. Porter read the manuscript and made numerous remarks which have been very helpful for its amendment, in particular, in the proper use of English. The benevolent, encouraging, and thorough comments on this work of Prof. K. Habetha have been thankfully incorporated. Prof. J. Ryan pointed out some important references, which are always essential in establishing the place of the book among the affine works. Prof. R. Gilbert's support was decisive for the book to have seen the light of day in the Pitman Research Notes in Mathematics Series.

We are indebted to all of them.

# Chapter 1

# $\alpha$-holomorphic function theory

## 1 Algebras of real and complex quaternions

**1.1** We shall denote by $\mathbb{H}(\mathbb{R})$ and by $\mathbb{H}(\mathbb{C})$ the sets of real and complex quaternions (=biquaternions) correspondingly (the letter $\mathbb{H}$ is chosen traditionally in honour of the inventor of quaternions, W.R. Hamilton).

This means that each quaternion $a$ is represented in the form $a = \sum_{k=0}^{3} a_k i_k$ where $\{a_k\} \subset \mathbb{R}$ for real quaternions and $\{a_k\} \subset \mathbb{C}$ for complex quaternions, $i_0$ is the unit and $\{i_k | k \in \mathbb{N}_3\}$ are the quaternionic imaginary units, that is, the standard basis elements possessing the following properties:

$$i_0^2 = i_0 = -i_k^2;\ i_0 i_k = i_k i_0 = i_k,\ k \in \mathbb{N}_3,$$

$$i_1 i_2 = -i_2 i_1 = i_3;\ i_2 i_3 = -i_3 i_2 = i_1;\ i_3 i_1 = -i_1 i_3 = i_2; \tag{1.1}$$

here and always $\mathbb{N}_p := \{1, 2, \ldots, p\}, \mathbb{N}_p^0 := \mathbb{N}_p \cup \{0\}$.

We denote the imaginary unit in $\mathbb{C}$ by $i$ as usual. By definition

$$i \cdot i_k = i_k \cdot i,\ k \in \mathbb{N}_p^0. \tag{1.2}$$

Mostly we will consider the set $\mathbb{H}(\mathbb{C})$ of complex quaternions which by the multiplication laws (1.1), (1.2) and by the natural component-wise operation of addition turns into a complex non-commutative, associative algebra with zero divisors. Of course the same operations turn $\mathbb{H}(\mathbb{R})$ into a real, non-commutative, associative algebra without zero divisors.

The set of complex quaternions is isomorphic as a real vectorial space to the set of octonions (Cayley numbers). So the difference between the sets lies on the algebraic level. By the definition of octonions the additional imaginary unit $i$ anticommutes with $i_k, k \in \mathbb{N}_3$ and, as a consequence, the algebra of octonions does not enjoy the property of associativity against the algebra of complex quaternions (see, e.g., [104, 55, 31]).

**1.2** Quaternions defined above as linear (real or complex) combinations of the units $i_k$ have many representations in other algebras, useful for applications. First of all if $a \in \mathbb{H}(\mathbb{C})$ then the vector representation $a = a_0 + \vec{a}$, where $\vec{a} := \sum_{k=1}^{3} a_k i_k$, will often be useful. $a_0$ will be called the scalar part and $\vec{a}$ the vector part of the complex quaternion $a : a_0 =: \mathrm{Sc}(a), \vec{a} =: \mathrm{Vec}(a)$.

A complex quaternion of the form $a = \vec{a}$ will be called purely vectorial. We identify them with vectors from $\mathbb{C}^3$. Replacing $\mathbb{H}(\mathbb{C})$ by $\mathbb{H}(\mathbb{R})$ we conserve the same terms with small obvious changes. For instance, $\vec{a} \in \mathbb{H}(\mathbb{R})$ is identified with a vector from $\mathbb{R}^3$.

In vector terms, the multiplication of two arbitrary complex quaternions $a, b$ can be rewritten as follows:

$$ab = a_0 b_0 - < \vec{a}, \vec{b} > + \left[\vec{a} \times \vec{b}\right] + a_0 \vec{b} + b_0 \vec{a}, \tag{1.3}$$

where

$$< \vec{a}, \vec{b} > := \sum_{k=1}^{3} a_k b_k \in \mathbb{C}, \tag{1.4}$$

$$[\vec{a} \times \vec{b}] := \begin{vmatrix} i_1 & i_2 & i_3 \\ a_1 & a_2 & a_3 \\ b_1 & b_2 & b_3 \end{vmatrix} \in \mathbb{C}^3, \tag{1.5}$$

will be called respectively the scalar product and the vector product of complex three-dimensional vectors.

**1.3** There are several representations of quaternions in matrix form. Let $b \in \mathbb{H}(\mathbb{C})$, and introduce the following matrices:

$$B_l(b) := \begin{pmatrix} b_0 & -b_1 & -b_2 & -b_3 \\ b_1 & b_0 & -b_3 & b_2 \\ b_2 & b_3 & b_0 & -b_1 \\ b_3 & -b_2 & b_1 & b_0 \end{pmatrix}, \tag{1.6}$$

$$B_r(b) := \begin{pmatrix} b_0 & -b_1 & -b_2 & -b_3 \\ b_1 & b_0 & b_3 & -b_2 \\ b_2 & -b_3 & b_0 & b_1 \\ b_3 & b_2 & -b_1 & b_0 \end{pmatrix}. \tag{1.7}$$

The matrix subalgebra $\mathcal{B}_l(\mathbb{C}) = \{B_l(b) \,|\, b \in \mathbb{H}(\mathbb{C})\}$ of $\mathbb{C}^{4\times 4}$ and $\mathbb{H}(\mathbb{C})$ are isomorphic as complex algebras. The same holds for $\mathcal{B}_r(\mathbb{C}) := \{B_r(b) \,|\, b \in \mathbb{H}(\mathbb{C})\}$ and $\mathbb{H}(\mathbb{C})$. This implies immediately that each of the matrix subalgebras $\mathcal{B}_l(\mathbb{R}) := \{B_l(b) \,|\, b \in \mathbb{H}(\mathbb{R})\}$ and $\mathcal{B}_r(\mathbb{R}) := \{B_l(b) \,|\, b \in \mathbb{H}(\mathbb{R})\}$ is isomorphic to $\mathbb{H}(\mathbb{R})$.

**1.4** Consider now the following standard Dirac matrices:

$$\gamma_0 := \begin{pmatrix} 1 & 0 & 0 & 0 \\ 0 & 1 & 0 & 0 \\ 0 & 0 & -1 & 0 \\ 0 & 0 & 0 & -1 \end{pmatrix}, \qquad \gamma_1 := \begin{pmatrix} 0 & 0 & 0 & -1 \\ 0 & 0 & -1 & 0 \\ 0 & 1 & 0 & 0 \\ 1 & 0 & 0 & 0 \end{pmatrix},$$

$$\gamma_2 := \begin{pmatrix} 0 & 0 & 0 & i \\ 0 & 0 & -i & 0 \\ 0 & -i & 0 & 0 \\ i & 0 & 0 & 0 \end{pmatrix}, \qquad \gamma_3 := \begin{pmatrix} 0 & 0 & -1 & 0 \\ 0 & 0 & 0 & 1 \\ 1 & 0 & 0 & 0 \\ 0 & -1 & 0 & 0 \end{pmatrix}. \tag{1.8}$$

They have the following well-known properties:

$$\gamma_0^2 = E_4, \text{ the identity matrix;}$$

$$\gamma_k^2 = -E_4, \ k \in \mathbb{N}_3;$$

$$\gamma_j\gamma_k + \gamma_k\gamma_j = 0 \text{ for } j, k \in \mathbb{N}_3^0.$$

Introduce the notation

$$\hat{i} := I; \ \hat{i}_1 := \gamma_3\gamma_2; \ \hat{i}_2 := \gamma_1\gamma_3; \hat{i}_3 := \gamma_1\gamma_2; \ \hat{i} := \gamma_0\gamma_1\gamma_2\gamma_3. \tag{1.9}$$

Simple calculations show that the set $\{\hat{i}_0, \hat{i}_1, \hat{i}_2, \hat{i}_3, \hat{i}\}$ of $4 \times 4$ complex matrices has the same multiplication laws as the set $\{i_0, i_1, i_2, i_3, i\}$. For example,

$$\begin{aligned} \hat{i}_1\hat{i}_2 &= \gamma_3\gamma_2\gamma_1\gamma_3 = -\gamma_2\gamma_1 = \hat{i}_3, \\ \hat{i}\hat{i}_1 &= \gamma_0\gamma_1\gamma_2\gamma_3\gamma_3\gamma_2 = \gamma_0\gamma_1 = \gamma_3\gamma_2\gamma_0\gamma_1\gamma_2\gamma_3 = \hat{i}_1\hat{i}. \end{aligned} \tag{1.10}$$

Denote by $\mathfrak{D}$ the (complex) algebra generated by $\left\{\hat{i}_0, \hat{i}_1, \hat{i}_2, \hat{i}_3, \hat{i}\right\}$, and define the map

$$\kappa : \mathbb{H}(\mathbb{C}) \longrightarrow \mathfrak{D}$$

by its action on the generators:

$$\kappa(i_k) =: \hat{i}_k, \kappa(i) =: \hat{i}. \tag{1.11}$$

It is clear that $\kappa$ is an isomorphism of complex algebras; hence $\mathfrak{D}$ is a matrix realization of $\mathbb{H}(\mathbb{C})$. If $a = \sum_{k=0}^{3} a_k \cdot i_k \in \mathbb{H}(\mathbb{C})$, $a_k = \tilde{a}_k + \tilde{\tilde{a}}_k \cdot i$, $\left\{\tilde{a}_k, \tilde{\tilde{a}}_k\right\} \subset \mathbb{R}$, then

$$\kappa(a) = \tilde{a}_0\hat{i}_0 + \tilde{a}_1 \cdot \hat{i}_1 + \tilde{a}_2 \cdot \hat{i}_2 + \tilde{a}_3\hat{i}_3 +$$

$$+\hat{i}\left(\tilde{\tilde{a}}_0\hat{i}_0 + \tilde{\tilde{a}}_1\hat{i}_1 + \tilde{\tilde{a}}_2\hat{i}_2 + \tilde{\tilde{a}}_3\hat{i}_3\right), \tag{1.12}$$

or, in explicit form,

$$\kappa(a) = \begin{pmatrix} \tilde{a}_0 - i\tilde{a}_3; & i\tilde{a}_1 + \tilde{a}_2; & i\tilde{\tilde{a}}_0 + \tilde{\tilde{a}}_3; & -\tilde{\tilde{a}}_1 + i\tilde{\tilde{a}}_2 \\ i\tilde{a}_1 - \tilde{a}_2; & \tilde{a}_0 + i\tilde{a}_3; & -\tilde{\tilde{a}}_1 - i\tilde{\tilde{a}}_2; & i\tilde{\tilde{a}}_0 - \tilde{\tilde{a}}_3 \\ i\tilde{\tilde{a}}_0 + \tilde{\tilde{a}}_3; & -\tilde{\tilde{a}}_1 + i\tilde{\tilde{a}}_2; & \tilde{a}_0 - i\tilde{a}_3; & i\tilde{a}_1 + \tilde{a}_2 \\ -\tilde{\tilde{a}}_1 - i\tilde{\tilde{a}}_2; & i\tilde{\tilde{a}}_0 - \tilde{\tilde{a}}_3; & i\tilde{a}_1 - \tilde{a}_2; & \tilde{a}_0 + i\tilde{a}_3 \end{pmatrix}. \tag{1.13}$$

This means that an arbitrary matrix $A \in \mathfrak{D}$ has the following form,

$$A = \begin{pmatrix} c_0 & -c_1^* & c_2 & c_3^* \\ c_1 & c_0^* & c_3 & -c_2^* \\ c_2 & c_3^* & c_0 & -c_1^* \\ c_3 & -c_2^* & c_1 & c_0^* \end{pmatrix}, \tag{1.14}$$

where $c_0, c_1, c_2, c_3$ are arbitrary complex numbers, and "$*$" denotes the usual complex conjugation.

**1.5** We will be in need of two different conjugations in $\mathbb{H}(\mathbb{C})$. For $a \in \mathbb{H}(\mathbb{C})$, the usual complex conjugation is defined by

$$Z_{\mathbb{C}}(a) := a^* = \operatorname{Re} a - i \operatorname{Im} a = \sum_{k=0}^{3} \operatorname{Re}(a_k) i_k - i \sum_{k=0}^{3} \operatorname{Im}(a_k) i_k,$$

and the quaternionic conjugation by

$$Z_{\mathbb{H}}(a) := \bar{a} = a_0 - \vec{a}.$$

If $\{a, b\} \subset \mathbb{H}(\mathbb{C})$ then

$$\overline{a \cdot b} = \bar{b} \cdot \bar{a}, \tag{1.15}$$

and if $a = \text{Re}\, a + i\, \text{Im}\, a$, then

$$a \cdot \bar{a} = \bar{a} \cdot a = \sum_{k=0}^{3} a_k^2 = |\text{Re}\, a|^2 - |\text{Im}\, a|^2 + 2i < \text{Re}\, a, \text{Im}\, a >_{\mathbb{R}^4} \in \mathbb{C}, \qquad (1.16)$$

where $|\text{Re}\, a|$, $|\text{Im}\, a|$ stand for the usual modulus of a real quaternion (or the Euclidean norm of a four-dimensional vector); $< \text{Re}\, a, \text{Im}\, a >_{\mathbb{R}^4}$ the Euclidean scalar product of two four-dimensional vectors.

(1.16) shows that the product $a\bar{a}$ can be zero even when $a$ is not, and hence, $\mathbb{H}(\mathbb{C})$ has zero divisors.

**1.6** Let us denote by $\mathfrak{S}$ the set of zero divisors from $\mathbb{H}(\mathbb{C})$; that is, $\mathfrak{S} := \{a \in \mathbb{H}(\mathbb{C}) \mid a \neq 0; \exists b \neq 0 : ab = 0\}$. Denote by $G\mathbb{H}(\mathbb{C})$ the subset of invertible elements from $\mathbb{H}(\mathbb{C})$. If $a \notin \mathfrak{S} \cup \{0\}$ then

$$a^{-1} := \bar{a}/(a\bar{a}) \qquad (1.17)$$

is the inverse of the complex quaternion $a$. Obviously, $G\mathbb{H}(\mathbb{C}) = \mathbb{H}(\mathbb{C}) \backslash (\mathfrak{S} \cup \{0\})$.

The following lemma gives some equivalent definitions of the biquaternionic zero divisors.

**1.7 Lemma** (Structure of the set of zero divisors) *Let* $0 \neq a \in \mathbb{H}(\mathbb{C})$. *The following conditions are equivalent:*

1. $a \in \mathfrak{S}$.

2. $a\bar{a} = 0$.

3. $a_0^2 = \vec{a}^2$.

4. $a^2 = 2a_0 a = 2\vec{a}a$.

**Proof.** First let us show the equivalence of 1) and 2). If 2) holds, then $b := \bar{a}$ satisfies the definition of $\mathfrak{S}$. If 2) does not hold, i.e., $a\bar{a} \neq 0$, then $\exists a^{-1}$ defined by (1.17) and $a \in G\mathbb{H}(\mathbb{C})$.

The equivalence of 2) and 3) follows immediately from the definition of quaternionic conjugation.

To prove the equivalence of 3) and 4) we have: $a^2 = a_0^2 + 2a_0\vec{a} + \vec{a}^2$. Hence $\vec{a}^2 = a_0^2 \Leftrightarrow a^2 = 2\vec{a}^2 + 2a_0\vec{a} = 2\vec{a}a$ and $\vec{a}^2 = a_0^2 \Leftrightarrow a^2 = 2a_0^2 + 2a_0\vec{a} = 2a_0 a$. □

# 2 Helmholtz operator with complex and quaternionic wave number

**2.1** We will consider $\mathbb{H}(\mathbb{C})$–valued functions given in a domain $\Omega \subset \mathbb{R}^3$. On the left $\mathbb{H}(\mathbb{C})$-module $C^2(\Omega; \mathbb{H}(\mathbb{C}))$ the following operator is defined:

$$\Delta_\lambda := \Delta + M^\lambda, \tag{2.1}$$

where $\Delta := \sum_{k=1}^{3} \partial_k^2$, $\partial_k := \frac{\partial}{\partial x_k}$; $M^\lambda[f] := f\lambda$, $\lambda \in \mathbb{H}(\mathbb{C})$. We will call the operator (2.1) the Helmholtz operator with a quaternionic wave number. When $\lambda = \lambda_0 \in \mathbb{C}$ we have, of course, the usual Helmholtz operator. When $\lambda \in \mathbb{H}(\mathbb{C})$, the operator (2.1) generates on the complex space $C^2(\Omega; \mathbb{C}^4)$ the following Helmholtz operator "with a matrix wave number" of a special form:

$$\Delta_{B_r(\lambda)} := \Delta \cdot E_4 + B_r(\lambda) \tag{2.2}$$

where $E_4$ is the identity matrix of the fourth order and $B_r(\lambda)$ is defined by (1.7).

Let us emphasize that even for the "non-exotic" case of $\lambda \in \mathbb{C}$, a great part of this section's results is new (see, for instance, Theorem 2.13 below on the decomposition of ker $\Delta_\lambda$). It should be mentioned that these "complex" (in the sense of "non-quaternionic") results were obtained only after the corresponding quaternionic tool had been elaborated.

It will be clear that all the following is true also for the operator ${}_\lambda\Delta := \Delta + \lambda I$, $\lambda \in \mathbb{H}(\mathbb{C})$, with obvious changes.

**2.2** On the set $C^1(\Omega; \mathbb{H}(\mathbb{C}))$ the well-known Moisil–Theodoresco operator is defined by the formula

$$D := \sum_{k=1}^{3} i_k \partial_k, \tag{2.3}$$

which was introduced for the first time in [87, 88] and then was investigated in an enormous number of works (see, e.g., [14, 132, 34, 43, 123, 46], etc.).

In vectorial language, the Moisil–Theodoresco system

$$D[f] = 0 \tag{2.4}$$

can be rewritten as

$$\operatorname{div}\vec{f} = 0, \tag{2.5}$$

$$\operatorname{grad} f_0 + \operatorname{rot}\vec{f} = 0. \tag{2.6}$$

The system (2.5)–(2.6) is considered by many as the most natural and simple spatial generalization of the Cauchy–Riemann equations.

A function $f : \Omega \subset \mathbb{R}^3 \longrightarrow \mathbb{H}(\mathbb{C})$ satisfying (2.4) in $\Omega$, i.e., $f \in \ker D(\Omega)$, will be called quaternionic holomorphic or hyperholomorphic in $\Omega$.

**2.3** From the definition of the operator $D$ its important property follows:

$$-D^2 = \Delta. \tag{2.7}$$

In particular, the equality (2.7) means that quaternionic holomorphic functions are necessarily harmonic, thus each property of harmonic functions has an immediate corollary for quaternionic holomorphic functions.

Examples include the mean value formula, the maximum modulus theorem and many others (see, e.g., [46], where it was demonstrated how to obtain these theorems for quaternionic holomorphic functions directly, without explicit application of (2.7)).

**2.4** Using matrix (1.6) the equality (2.4) can be also rewritten as

$$B_l\left(\sum_{k=1}^{3} i_k\partial_k\right) f = 0 \tag{2.8}$$

with

$$B_l\left(\sum_{k=1}^{3} i_k\partial_k\right) = \begin{pmatrix} 0 & -\partial_1 & -\partial_2 & -\partial_3 \\ \partial_1 & 0 & -\partial_3 & \partial_2 \\ \partial_2 & \partial_3 & 0 & -\partial_1 \\ \partial_3 & -\partial_2 & \partial_1 & 0 \end{pmatrix}. \tag{2.9}$$

Thus we have a partial differential operator of first order with constant coefficients factorizing the three-dimensional Laplace operator.

A natural question arises: What is the general form for a partial differential operator of first order factorizing the Laplace operator? Let us put this question more precisely.

In $n$-dimensional Euclidean space introduce a matrix differential operator

$$\mathcal{D} := \sum_{k=1}^{n} \sigma_k\partial_k, \tag{2.10}$$

where $\sigma_k$ are real constant matrices of order $p \times p$. Suppose that together with its conjugate

$$\mathcal{D}' := \sum_{k=1}^{n} \sigma_k^T\partial_k$$

(the index $T$ denotes transposition) the operator (2.10) satisfies the condition

$$\mathcal{D}'\mathcal{D} = E_p\Delta_n \tag{2.11}$$

where $\Delta_n$ is the $n$-dimensional Laplace operator and $E_p$ is the identity matrix of order $p$. Then what conditions must be placed on the matrices $\{\sigma_k\}$ such that the operator $\mathcal{D}$ satisfies the equality (2.11)?

The answer is well known. The relation (2.11) is equivalent to the following:

$$\sigma_k^T\sigma_j + \sigma_j^T\sigma_k = 2\delta_{kj}E_p, \qquad j,k \in \mathbb{N}_n, \tag{2.12}$$

where $\delta_{kj}$ is the Kronecker delta.

The conditions (2.12) impose severe constraints on the order $p$ of the square matrices $\sigma_k$. Namely, as was shown by W. A. Hurwitz [49] (see also [115]), $p$ must be divisible by some power of 2 depending on $n$. If $n$ is represented in the form $n = 8b + r$, where $1 \le r \le 8$ and $r \le 2^c$ then $p$ is divisible by $p_0 := 2^{4b+c}$. Consequently, $p = n$ iff $n = 2, 4, 8$.

In [17] a method was given for constructing (by induction) matrices $\sigma_k$ of order $2^{n-1}$. Other problems associated with the construction of appropriate matrices $\sigma_k$ and corresponding analogues of the Cauchy–Riemann and the Moisil–Theodoresco systems were studied in [30, 118, 36, 44, 129, 5, 6].

**2.5** For the case of quaternions, let us define on $C^1(\Omega; \mathbb{H}(\mathbb{C}))$ an operator ${}^{\psi}D$ by the formula

$${}^{\psi}D[f] := \sum_{k=1}^{3} \psi^k \partial_k f, \tag{2.13}$$

where $\psi := \{\psi^1, \psi^2, \psi^3\}$, $\psi^k \in \mathbb{H}(\mathbb{R})$. Denote $\bar{\psi} := \left\{\bar{\psi}^1, \bar{\psi}^2, \bar{\psi}^3\right\}$. It was demonstrated in [93, 126] that the equalities

$${}^{\psi}D\,{}^{\bar{\psi}}D = {}^{\bar{\psi}}D\,{}^{\psi}D = \Delta \tag{2.14}$$

are true iff

$$\psi^j\bar{\psi}^k + \psi^k\bar{\psi}^j = 2\delta_{jk} \tag{2.15}$$

for any $j, k$ from $\mathbb{N}_3$.

The condition (2.15) is obtained from (2.14) by a straightforward calculation. We shall call any set $\psi$ of real quaternions with the property (2.15) a "structural set". It is evident that $\psi$ and $\bar{\psi}$ are structural sets simultaneously.

Reasoning along similar lines we can consider the operator

$$D^{\psi}[f] := \sum_{k=1}^{3}(\partial_k f)\psi^k.$$

From (1.15) it follows that

$${}^{\psi}D = Z_{\mathbb{H}}D^{\bar{\psi}}Z_{\mathbb{H}}, \quad D^{\psi} = Z_{\mathbb{H}}{}^{\bar{\psi}}DZ_{\mathbb{H}}, \tag{2.16}$$

so it is sufficient to study the "left" (or the "right") case only; we confine the discussion to the left one.

Let us note that for purely imaginary $\psi$ the factorization (2.14) becomes

$$-{}^{\psi}D^2 = \Delta. \tag{2.17}$$

**2.6** Let us denote by $\alpha$ a complex-quaternionic square root of $\lambda$ appearing in (2.1). That is, $\alpha \in \mathbb{H}(\mathbb{C})$ and $\alpha^2 = \lambda$. Let $\psi := \{\psi^1, \psi^2, \psi^3\}$ consist of purely imaginary quaternions only: $\mathrm{Sc}(\psi^1) = \mathrm{Sc}(\psi^2) = \mathrm{Sc}(\psi^3) = 0$ (some remarks on this will be given below in Subsection 3.7). We will introduce a "metaharmonic generalization" of the operator (2.13). Let

$${}^{\psi}D_{\alpha} := {}^{\psi}D + M^{\alpha}.$$

Then from (2.17) the following factorization arises:

$$\Delta_{\lambda} = -{}^{\psi}D_{\alpha}{}^{\psi}D_{-\alpha} = -{}^{\psi}D_{-\alpha}{}^{\psi}D_{\alpha}. \tag{2.18}$$

Note that for $\lambda = \lambda_0 \in \mathbb{C}$ we have not only complex square roots $\pm\sqrt{\lambda_0}$ but also a family of purely imaginary complex quaternions $\vec{\alpha}$ satisfying the condition $-\alpha_1^2 - \alpha_2^2 - \alpha_3^2 = \lambda_0$.

Moreover, if $\lambda = 0$, i.e., for the Laplace operator, the equality (2.14) is not the only factorization. When $\alpha \in \mathfrak{S}$ and $\alpha_0 = 0$ we obtain

$$\Delta = -{}^{\psi}D_{\vec{\alpha}}{}^{\psi}D_{-\vec{\alpha}} = -{}^{\psi}D_{-\vec{\alpha}}{}^{\psi}D_{\vec{\alpha}}. \tag{2.19}$$

Thus a family is constructed of partial differential operators of first order factorizing the Helmholtz operator (2.1) and the "usual" Laplace operator.

**2.7** Consider now the usual complex one-dimensional Cauchy–Riemann operators

$$\bar{\partial} := \frac{1}{2}(\partial_x + i\partial_y), \qquad \partial := \frac{1}{2}(\partial_x - i\partial_y). \tag{2.20}$$

More precisely, now we are on the complex plane $\mathbb{C}$, $z = x + iy$, $z^* = x - iy, \partial_x = \frac{\partial}{\partial x}$, $\partial_y = \frac{\partial}{\partial y}$, $z \in \Omega_2$, a simply connected domain. Each of the operators ${}^{\psi}D, {}^{\psi}D_{\alpha}$ is a generalization of $\bar{\partial}$, and factorizations (2.7), (2.14), (2.17), (2.18), (2.19) are direct generalizations of the following property of the operators (2.20):

$$\partial \cdot \bar{\partial} = \frac{1}{4}\Delta_2 \tag{2.21}$$

with $\Delta_2$ the two-dimensional Laplace operator. It is known that

$$\ker\partial \bigcap \ker \bar{\partial} = \{const \ \in \mathbb{C}\} = \mathbb{C} \tag{2.22}$$

and that each complex harmonic function (i.e., a function $w = u + iv$ such that $\Delta_2[u] = \Delta_2[v] = 0$) can be represented as a sum $w = f + g$ for $f \in \ker \bar{\partial}$, $g \in \ker \partial$. This representation can be constructed explicitly. Indeed, for $u$ harmonic and real-valued let

$$u_1(z) := i\int_{z_0}^{z} \left(\bar{\partial}[u]d\bar{z} - \partial[u]dz\right), \tag{2.23}$$

then $u + iu_1$ is holomorphic in $\Omega$. The same may be done for $v$, and it is clear now what $f$ and $g$ are in $w = f + g$.

Equality (2.22) implies that the general representation for $w$ is

$$w = F + G \tag{2.24}$$

with $F = f + c$, $G = g - c$ for an arbitrary complex constant $c$. This means that the following decomposition of the space of complex-valued harmonic functions holds:

$$\ker\Delta_2 \approx (\ker\partial/\mathbb{C}) \bigoplus (\ker\bar{\partial}/\mathbb{C}) \bigoplus \mathbb{C} \tag{2.25}$$

where " $\approx$ " denotes isomorphism of linear complex spaces. The quotient space $\ker\bar{\partial}/\mathbb{C}$ can be understood as a space of holomorphic functions vanishing at a fixed point of $\Omega_2$ (analogously for $\ker\partial/\mathbb{C}$). In fact, let $z_0 \in \Omega_2$ be fixed. Denote:

$$\begin{aligned} \ker_{z_0}\partial &:= \{g \mid g \in \ker\partial, g(z_0) = 0\}, \\ \ker_{z_0}\bar{\partial} &:= \left\{f \mid f \in \ker\bar{\partial}, f(z_0) = 0\right\}. \end{aligned} \tag{2.26}$$

Then

$$\ker\Delta_2 = \ker_{z_0}\partial \bigoplus \ker_{z_0}\bar{\partial} \bigoplus \mathbb{C}. \tag{2.27}$$

It is easy to describe now the corresponding projectors. Denote them by $\Pi_{\partial,z_0}$, $\Pi_{\bar{\partial},z_0}$, $\Pi_{\mathbb{C},z_0}$. Then for $w = u + iv \in \ker\Delta_2$

$$\Pi_{\mathbb{C},z_0}[w] = w(z_0), \tag{2.28}$$

$$\Pi_{\bar{\partial},z_0}[w] = w + iw_1 - (w(z_0) + i\, w_1(z_0)) \tag{2.29}$$

where $w_1 = u_1 + iv_1$, $u_1$ is defined by (2.23), and $v_1$ analogously;

$$\Pi_{\partial,z_0}[w] = \overline{\Pi_{\bar{\partial},z_0}[w]}. \tag{2.30}$$

**2.8** Returning now to the multidimensional situation, we will see that although there are some nice analogues of the decompositions (2.25) and (2.27), there are many surprising peculiarities both in techniques and in final formulas. First we consider the case of $\lambda \neq 0$.

Note that if $0 \neq \lambda \in \mathfrak{S}$, then $\mathrm{Sc}(\alpha) \neq 0$ because $\vec{\alpha}^2 \notin \mathfrak{S}$ (see Lemma 1.7). Therefore, unless otherwise stipulated, we assume $\mathrm{Sc}(\alpha) \neq 0$ when $\alpha \in \mathfrak{S}$.

Introduce the operator

$$ {}^{\psi}\pi_\alpha := \begin{cases} -(2\alpha\bar{\alpha})^{-1} M^{\bar{\alpha}\,\psi} D_{-\alpha}, & \alpha \notin \mathfrak{S}, \\ \\ -(8\alpha_0^2)^{-1} M^{\alpha\,\psi} D_{-\alpha}, & \alpha \in \mathfrak{S}, \end{cases} \tag{2.31} $$

defined on $C^1(\Omega; \mathbb{H}(\mathbb{C}))$. We will consider its restriction ${}^{\psi}\Pi_\alpha := {}^{\psi}\pi_\alpha \mid \ker\Delta_\lambda$.

The operator ${}^{\psi}\Pi_\alpha$ acts invariantly on $\ker\Delta_\lambda$: if $f \in \ker\Delta_\lambda$ then

$$ \Delta_\lambda \left[ {}^{\psi}\Pi_\alpha[f] \right] = {}^{\psi}\Pi_\alpha \left[ \Delta_\lambda[f] \right] = 0. $$

This computation involves the commutativity of the multiplication of the operators $M^{\bar{\alpha}}$ and ${}^{\psi}D_{-\alpha}$:

$$ M^{\bar{\alpha}} \cdot {}^{\psi}D_{-\alpha} = {}^{\psi}D_{-\alpha} \cdot M^{\bar{\alpha}}. $$

Let us remark that for $\alpha \in \mathfrak{S}$, another representation for the operator ${}^{\psi}\Pi_\alpha$ is obtained from Lemma 1.7:

$$ {}^{\psi}\Pi_\alpha = -\left(8\alpha_0^2\right)^{-1} M^{\alpha} \cdot {}^{\psi}D_{-2\alpha_0}. \tag{2.32} $$

**2.9 Proposition** (Properties of the operator ${}^{\psi}\Pi_\alpha$) *The following relations hold:*

*1.* ${}^{\psi}\Pi_\alpha^2 = {}^{\psi}\Pi_\alpha, \ \forall \alpha \in \mathbb{H}(\mathbb{C})$,

*2.* ${}^{\psi}\Pi_\alpha {}^{\psi}\Pi_{-\alpha} = {}^{\psi}\Pi_{-\alpha} {}^{\psi}\Pi_\alpha = 0, \ \forall \alpha \in \mathbb{H}(\mathbb{C})$,

*3.* ${}^{\psi}\Pi_\alpha + {}^{\psi}\Pi_{-\alpha} = I$, $\forall \alpha \notin \mathfrak{S}$,

$$ {}^{\psi}\Pi_\alpha + {}^{\psi}\Pi_{-\alpha} + (2\alpha_0)^{-1} M^{\bar{\alpha}} = I \tag{2.33} $$

*and*

$$ {}^{\psi}\Pi_\alpha M^{\bar{\alpha}} = M^{\bar{\alpha}} {}^{\psi}\Pi_\alpha = 0, \ \forall \alpha \in \mathfrak{S}, \tag{2.34} $$

*4.* $$ \operatorname{im} {}^{\psi}\Pi_\alpha = \begin{cases} \ker({}^{\psi}\pi_{-\alpha} \mid C^2), & \alpha \notin \mathfrak{S}, \\ \ker({}^{\psi}\pi_{-\alpha} \mid C^2) \bigcap \ker(M^{\bar{\alpha}} \mid C^2), & \alpha \in \mathfrak{S}. \end{cases} $$

PROOF.

1. First, let $\alpha \notin \mathfrak{S}$. For $f \in \ker\Delta_\lambda$ taking into account equality (2.18) we see:

$$ {}^{\psi}D^2_{-\alpha}[f] = \left(-M^\alpha + {}^{\psi}D\right)^2 [f] = \left(M^{\alpha^2} - 2M^{\alpha\psi}D - \Delta\right)[f] = $$

$$ = 2M^\alpha \left(M^\alpha - {}^{\psi}D\right)[f] = -2M^\alpha \cdot {}^{\psi}D_{-\alpha}[f]. $$

Therefore

$$ {}^{\psi}\Pi^2_\alpha[f] = (2\alpha\bar{\alpha})^{-2} M^{\bar{\alpha}^2} \cdot {}^{\psi}D^2_{-\alpha}[f] = $$

$$ = (2\alpha\bar{\alpha})^{-2} M^{\bar{\alpha}^2} \left(-2M^\alpha \cdot {}^{\psi}D_{-\alpha}\right)[f] = {}^{\psi}\Pi_\alpha[f]. $$

When $\alpha \in \mathfrak{S}$ using Lemma 1.7 we obtain for a function $f \in \ker\Delta_\lambda$:

$$ {}^{\psi}\Pi^2_\alpha[f] = \left(8\alpha_0^2\right)^{-2} M^\alpha \cdot {}^{\psi}D_{-\alpha}\left[{}^{\psi}D[f]\alpha - 2\alpha_0 f\alpha\right] = $$

$$ = \left(8\alpha_0^2\right)^{-2} M^\alpha \left(-{}^{\psi}D[f]\alpha^2 + 2\alpha_0 f\alpha^2 - \Delta[f]\alpha - 2\alpha_0 {}^{\psi}D[f]\alpha\right) = $$

$$= \left(8\alpha_0^2\right)^{-2} M^\alpha \left(-4\alpha_0 {}^\psi D[f]\alpha + 8\alpha_0^2 f\alpha\right) =$$

$$= \left(8\alpha_0^2\right)^{-2} M^\alpha \left(-8\alpha_0^2 \cdot {}^\psi D[f] + 8\alpha_0^2 M^\alpha[f]\right) =$$

$$= \left(8\alpha_0^2\right)^{-1} M^\alpha \left(-{}^\psi D[f] + M^\alpha f\right) = {}^\psi \Pi_\alpha[f].$$

2. Is a direct consequence of the factorization (2.18) and of the evident fact that the operator ${}^\psi D_{-\alpha}$ commutes with the operators $M^\alpha$ and $M^{\bar\alpha}$.

3. Follows directly from the definition of the operator ${}^\psi\Pi_\alpha$. In order to obtain the equalities (2.34) one should also recall assertion 2) of Lemma 1.7.

4. First, let $\alpha \notin \mathfrak{S}$ and $f \in \operatorname{im}{}^\psi\Pi_\alpha$, i.e., $\exists g \in \ker \Delta_\lambda$ such that $f = {}^\psi\Pi_\alpha[g]$. Then

$${}^\psi\pi_{-\alpha}[f] = {}^\psi\pi_{-\alpha}\left[{}^\psi\Pi_\alpha[g]\right] = {}^\psi\pi_{-\alpha}[{}^\psi\pi_\alpha[g]] = (2\alpha\bar\alpha)^{-2} M^{\bar\alpha^2} \Delta_\lambda[g] = 0.$$

Hence, $\operatorname{im}{}^\psi\Pi_\alpha \subset \ker\left({}^\psi\pi_{-\alpha} \mid C^2\right)$.

If $f \in \ker\left({}^\psi\pi_{-\alpha} \mid C^2\right)$, then applying the operator ${}^\psi D_{-\alpha}$ to the equality ${}^\psi\pi_{-\alpha}[f] = 0$ we obtain $f \in \ker\Delta_\lambda$. Consequently, $\ker\left({}^\psi\pi_{-\alpha} \mid C^2\right) \subset \ker\Delta_\lambda$ and $f \in \ker{}^\psi\Pi_{-\alpha}$. Then from assertion 3 of this proposition we see that

$$f = {}^\psi \Pi_\alpha[f] \in \operatorname{im}{}^\psi\Pi_\alpha.$$

It remains to prove 4) when $\alpha \in \mathfrak{S}$. The inclusion $\operatorname{im}{}^\psi\Pi_\alpha \subset \ker\left({}^\psi\pi_{-\alpha} \mid C^2\right)$ can be proved by analogy with the above reasoning. Further, if $f \in \operatorname{im}{}^\psi\Pi_\alpha$, then $M^{\bar\alpha}[f] = 0$ (see assertion 2 of Lemma 1.7). Hence

$$\operatorname{im}{}^\psi\Pi_\alpha \subset \ker\left({}^\psi\pi_{-\alpha} \mid C^2\right) \bigcap \ker\left(M^{\bar\alpha} \mid C^2\right).$$

Conversely, if $f \in \ker\left({}^{\psi}\pi_{-\alpha} \mid C^2\right) \bigcap \ker\left(M^{\bar{\alpha}} \mid C^2\right)$ then

$$M^{\alpha} \cdot {}^{\psi}D_{-\alpha}[f] = 0.$$

Consequently, $\Delta_{\lambda}[f\alpha] = 0$ also.

Further, $f\bar{\alpha} = 0$ and we have

$$\Delta_{\lambda}[f\alpha + f\bar{\alpha}] = 0,$$

from which it follows that $2\alpha_0\Delta_{\lambda}[f] = 0$. That is, $f \in \ker\Delta_{\lambda}$. Therefore $f \in \ker{}^{\psi}\Pi_{-\alpha} \bigcap \ker M^{\bar{\alpha}}$. Then from assertion 3 of this proposition it follows again that

$$f \in \operatorname{im}{}^{\psi}\Pi_{\alpha}.$$

□

**2.10** The previous proposition shows that the operators ${}^{\psi}\Pi_{\alpha}, {}^{\psi}\Pi_{-\alpha}$ are orthogonal projectors on $\ker\Delta_{\lambda}$. Moreover, in the case of $\alpha \notin \mathfrak{S}$ they are mutually complementary projectors. When $\alpha \in \mathfrak{S}$ the situation is more sophisticated. They are not mutually complementary, but introducing the operator $(2\alpha_0)^{-1} M^{\alpha}$ orthogonal to ${}^{\psi}\Pi_{\pm\alpha}$ we obtain as a sum the identity operator (equality (2.33)). The following proposition describes the action of this new operator $(2\alpha_0)^{-1}M^{\alpha}$.

**2.11 Proposition** (Multiplication by zero divisors and projectors). *Let $\alpha \in \mathfrak{S}$. Then the operators $(2\alpha_0)^{-1} M^{\alpha}$ and $(2\alpha_0)^{-1} M^{\bar{\alpha}}$ are mutually complementary projectors, and*

$$\operatorname{im}\left(M^{\alpha} \mid \ker\Delta_{\lambda}\right) = M^{\alpha}\left(\ker\Delta_{2\lambda_0}\right), \tag{2.35}$$

$$\operatorname{im}(M^{\bar{\alpha}} \mid \ker\Delta_\lambda) = M^{\bar{\alpha}}(\ker\Delta). \tag{2.36}$$

PROOF. That the operators $(2\alpha_0)^{-1}M^{\alpha}$ and $(2\alpha_0)^{-1}M^{\bar{\alpha}}$ are mutually complementary projectors follows from assertion 2) of Lemma 1.7 and from the equality $\alpha + \bar{\alpha} = 2\alpha_0$.

Let $f \in \ker\Delta_\lambda$. Then

$$\Delta_{2\lambda_0}[f\alpha] = \left(\Delta[f] + f\left(\lambda + \bar{\lambda}\right)\right)\alpha = f\bar{\lambda}\alpha =$$

$$= f\bar{\alpha}^2\alpha = 0,$$

since $\alpha \in \mathfrak{S}$.

By analogy, from the assumption $f \in \ker\Delta_\lambda$, we get

$$\Delta\left[f\bar{\alpha}\right] = -f\lambda\bar{\alpha} = 0.$$

Thus

$$\operatorname{im}(M^{\alpha} \mid \ker\Delta_\lambda) \subset M^{\alpha}(\ker\Delta_{2\lambda_0}),$$

$$\operatorname{im}(M^{\bar{\alpha}} \mid \ker\Delta_\lambda) \subset M^{\bar{\alpha}}(\ker\Delta).$$

On the other hand, if $g \in \ker\Delta_{2\lambda_0}$ then from

$$\Delta_\lambda[g\alpha] = -2\lambda_0 g\alpha + g\lambda\alpha = -g\bar{\lambda}\alpha = 0$$

we get $M^{\alpha}[g] \in \ker\Delta_\lambda$.

By analogy, $M^{\bar{\alpha}}[h] \in \ker\Delta_\lambda$ when $h \in \ker\Delta$.

Consequently,

$$M^{\alpha}(\ker\Delta_{2\lambda_0}) \subset \operatorname{im}(M^{\alpha} \mid \ker\Delta_\lambda)$$

and

$$M^{\bar{\alpha}}(\ker\Delta) \subset \operatorname{im}(M^{\bar{\alpha}} \mid \ker\Delta_\lambda).$$

□

This proposition implies the following corollary.

**2.12 Corollary** *For* $\alpha \in \mathfrak{S}$,

$$^{\psi}\Pi_\alpha : \ker\Delta_\lambda \longrightarrow \ker\Delta_{2\lambda_0}.$$

PROOF. This follows from the representation of $^{\psi}\Pi_\alpha$ (2.32) and from Proposition 2.11. □

Now everything is ready to prove the very important and useful assertion on the decomposition of the null set of the Helmholtz operator with a quaternionic wave number.

Functions from $\ker\Delta_\lambda$ will be called metaharmonic (or more precisely, $\lambda$ metaharmonic). It is clear that 0-metaharmonic functions are nothing more than harmonic. We shall sometimes use the notation $\mathcal{H}_\lambda(\Omega, \mathbb{H}(\mathbb{C}))$ or $\mathcal{H}_\lambda$ for the left $\mathbb{H}(\mathbb{C})$-module $\ker\Delta_\lambda$ comprising all metaharmonic functions. Instead of $\mathcal{H}_0$ we shall write $\mathcal{H}$.

**2.13 Theorem** (Decomposition of the space of metaharmonic functions). *Let* $\lambda \neq 0$. *If* $\lambda \notin \mathfrak{S}$, *then*

$$\ker\Delta_\lambda = \ker{}^{\psi}D_\alpha \bigoplus \ker{}^{\psi}D_{-\alpha}. \tag{2.37}$$

*If* $\lambda \in \mathfrak{S}$, *then*

$$\ker\Delta_\lambda = M^{\alpha}\left(\ker{}^{\psi}D_{2\alpha_0}\right) \bigoplus M^{\alpha}\left(\ker{}^{\psi}D_{-2\alpha_0}\right) \bigoplus M^{\bar{\alpha}}(\ker\Delta) =$$

$$= \left(\ker{}^{\psi}D_{2\alpha_0}\right)\cdot\alpha \bigoplus \left(\ker{}^{\psi}D_{-2\alpha_0}\right)\cdot\alpha \bigoplus (\ker\Delta)\,\bar{\alpha}. \tag{2.38}$$

PROOF. First, let $\lambda \notin \mathfrak{S}$. Note that

$$\mathrm{im}^{\psi}\Pi_{\alpha} = \ker{}^{\psi}D_{\alpha}.$$

Then the equality (2.37) follows directly from Proposition 2.9.

Let us consider the case $\lambda \in \mathfrak{S}$. From Proposition 2.11 we see that

$$\ker\Delta_{\lambda} = M^{\alpha}\left(\ker\Delta_{2\lambda_0}\right) \bigoplus M^{\bar{\alpha}}(\ker\Delta).$$

It is clear that $2\lambda_0 \notin \mathfrak{S}$, and we can use equality (2.37) for $\ker\Delta_{2\lambda_0}$ (taking into account that $2\lambda_0 = 4\alpha_0^2$ by Lemma 1.7). We have

$$\ker\Delta_{\lambda} = M^{\alpha}\left(\ker{}^{\psi}D_{2\alpha_0} \bigoplus \ker{}^{\psi}D_{-2\alpha_0}\right) \bigoplus M^{\bar{\alpha}}(\ker\Delta),$$

from which (2.38) follows. □

**2.14** Of course, each of the equalities (2.37) and (2.38) is a good analogue of the equalities (2.25) and (2.27). (2.37) takes into account that $\ker{}^{\psi}D_{\alpha} \bigcap \ker{}^{\psi}D_{-\alpha} = \{0\}$ for $\lambda \notin \mathfrak{S}$, while (2.38) includes other peculiarities of the case, for instance: if $\lambda$ is a zero divisor, then any $\lambda$-metaharmonic function is a "direct sum" of three functions, the last of them being harmonic!

The proof is based strongly on the fact that $\alpha \neq 0$, and it becomes invalid for $\alpha = 0$. But the essence of the matter is more profound, and it is impossible in principle to prove (2.37) for $\alpha = 0$. In other words, the space of harmonic functions of three variables has no direct decomposition by $\ker{}^{\psi}D$.

More interestingly, such a decomposition can be constructed using again the operators ${}^{\psi}D_{\alpha}$ with $\alpha$ zero divisors.

Let $\alpha \in \mathfrak{S}$ and suppose $\alpha_0 = 0$. Then the equality (2.19) holds and, further, $\ker{}^{\psi}D_{\pm\alpha} \subset \ker\Delta$. But the null set of the Laplace operator is not representable as a direct sum of the null sets of the operators ${}^{\psi}D_{\pm\alpha}$, because for such $\alpha$, $\ker{}^{\psi}D_{\alpha} \bigcap \ker{}^{\psi}D_{-\alpha} \neq \emptyset$. In fact, we have the following

**2.15 Proposition** *For* $\alpha \in \mathfrak{S}$, $\alpha_0 = 0$,

$$\ker{}^{\psi}D_{\alpha} \bigcap \ker{}^{\psi}D_{-\alpha} = M^{\alpha}\left(\ker{}^{\psi}D\right).$$

PROOF. First let us show that

$$M^{\alpha}\left(\ker{}^{\psi}D\right) = \ker{}^{\psi}D \bigcap \ker M^{\alpha}. \tag{2.39}$$

We can write

$$M^{\alpha}(\ker{}^{\psi}D) = \ker{}^{\psi}D \bigcap \operatorname{im} M^{\alpha}.$$

But from the assumption on $\alpha$ it follows that

$$\operatorname{im} M^{\alpha} = \ker M^{\alpha}$$

and we obtain (2.39).

It remains to show the simple equality

$$\ker{}^{\psi}D_{\alpha} \bigcap \ker{}^{\psi}D_{-\alpha} = \ker{}^{\psi}D \bigcap \ker M^{\alpha}. \tag{2.40}$$

The condition $f \in \ker{}^{\psi}D_{\alpha} \bigcap \ker{}^{\psi}D_{-\alpha}$ signifies that ${}^{\psi}D[f] = -f\alpha$ and ${}^{\psi}D[f] = f\alpha$ simultaneously. Adding and subtracting the last two equalities we see ${}^{\psi}D[f] = 0$ and $f\alpha = 0$ simultaneously. Thus we obtain (2.40). □

**2.16** As before, we assume that $\alpha \in \mathfrak{S}$, $\alpha_0 = 0$. At least one of the numbers $\alpha_1, \alpha_2, \alpha_3$ differs from zero; to be definite, assume that $\alpha_1 \neq 0$. On $\ker\Delta$ let us introduce a pair of operators ${}^{\psi}\Pi_{\alpha}^{\pm}$ :

$$ {}^{\psi}\Pi_{\alpha}^{+}[f] := -(2\alpha_1)^{-1\,\psi} D_{\alpha} M^{i_1}[f]; \tag{2.41}$$

$$ {}^{\psi}\Pi_{\alpha}^{-}[f] := \left(I - {}^{\psi}\Pi_{\alpha}^{+}\right)[f] = (2\alpha_1)^{-1}\left\{{}^{\psi}D[f]i_1 + f\,(2\alpha_1 + i_1\alpha)\right\} =$$

$$= (2\alpha_1)^{-1} \left({}^{\psi}DM^{i_1} - M^{i_1}M^{\alpha}\right)[f] = (2\alpha_1)^{-1} M^{i_1\psi}D_{-\alpha}[f].$$

The following proposition demonstrates that they are mutually orthogonal projectors.

**2.17 Proposition** *The following equalities hold:*

1. $\left({}^{\psi}\Pi_{\alpha}^{\pm}\right)^2 = {}^{\psi}\Pi_{\alpha}^{\pm}$,

2. ${}^{\psi}\Pi_{\alpha}^{+} \cdot {}^{\psi}\Pi_{\alpha}^{-} = {}^{\psi}\Pi_{\alpha}^{-} \cdot {}^{\psi}\Pi_{\alpha}^{+} = 0$.

PROOF. It is sufficient to verify that $\left({}^{\psi}\Pi_{\alpha}^{+}\right)^2 = {}^{\psi}\Pi_{\alpha}^{+}$.

Let us consider

$$\left({}^{\psi}\Pi_{\alpha}^{+}\right)^2[f] = (2\alpha_1)^{-2}\left(f\,(i_1\alpha)^2 + M^{\alpha}M^{i_1\psi}D\,[fi_1] + \right.$$

$$\left. + {}^{\psi}DM^{i_1}\,[fi_1\alpha]\right).$$

Here we have used the inclusion $f \in \ker\Delta$. Note that $(i_1\alpha) \in \mathfrak{S}$. Therefore, using the assertion 4) of Lemma 1.7 we see

$$(i_1\alpha)^2 = 2\mathrm{Sc}\,(i_1\alpha)\,i_1\alpha = -2\alpha_1 i_1\alpha.$$

Besides, $-\alpha + i_1\alpha i_1 = -2\alpha_1 i_1$. Hence

$$M^{\alpha}M^{i_1\psi}D\,[fi_1] + {}^{\psi}DM^{i_1}\,[fi_1\alpha] = -2\alpha_1\, M^{i_1\psi}D[f],$$

and we obtain

$$\left({}^{\psi}\Pi_{\alpha}^{+}\right)^2[f] = -(2\alpha_1)^{-1}\left(M^{i_1\alpha}[f] + M^{i_1}\,{}^{\psi}D[f]\right) = {}^{\psi}\Pi_{\alpha}^{+}[f].$$

□

Naturally this proposition leads to a decomposition of the harmonic function space.

**2.18 Theorem** (Decomposition of the space of harmonic functions). *Let* $\alpha \in \mathfrak{S}$, $\alpha_0 = 0, \alpha_1 \neq 0$.

*Then*

$$\ker\Delta = \ker{}^{\psi}D_{-\alpha} \bigoplus M^{i_1}\left(\ker{}^{\psi}D_{\alpha}\right). \tag{2.42}$$

PROOF. By virtue of Proposition 2.17 this theorem will be completely proved when we have shown the equalities

$$\operatorname{im}{}^{\psi}\Pi^{+}_{\alpha} = \ker{}^{\psi}D_{-\alpha}, \quad \ker{}^{\psi}\Pi^{+}_{\alpha} = M^{i_1}\left(\ker{}^{\psi}D_{\alpha}\right) \tag{2.43}$$

hold.

To prove (2.43) let us consider a function $f \in \ker{}^{\psi}D_{-\alpha}$. Then

$${}^{\psi}\Pi^{+}_{\alpha}[f] = -(2\alpha_1)^{-1}\left({}^{\psi}D[f]i_1 + fi_1\alpha\right) = -(2\alpha_1)^{-1}f\left(\alpha i_1 + i_1\alpha\right),$$

from which we have

$${}^{\psi}\Pi^{+}_{\alpha}[f] = [f],$$

since

$$\alpha_1 i_1 + i_1\alpha = -i_1\left(-\alpha + i_1\alpha i_1\right) = -2\alpha_1.$$

Consequently, $\ker{}^{\psi}D_{-\alpha} \subset \operatorname{im}{}^{\psi}\Pi^{+}_{\alpha}$.

When $f \in \operatorname{im}{}^{\psi}\Pi^{+}_{\alpha}$, that is, $\exists g \in \ker\Delta$ such that $f = {}^{\psi}\Pi^{+}_{\alpha}[g]$, we have

$${}^{\psi}D_{-\alpha}[f] = -(2\alpha_1)^{-1}\Delta\left[g\, i_1\right] = 0.$$

Thus the first of the equalities (2.43) is obtained.

Let $f \in \ker{}^{\psi}\Pi^{+}_{\alpha}$. From the definition of ${}^{\psi}\Pi^{+}_{\alpha}$ it follows that $f \in \ker{}^{\psi}\Pi^{+}_{\alpha}$ is equivalent to the inclusion $(fi_1) \in \ker{}^{\psi}D_{\alpha}$. Let us show that this is equivalent to

$$f \in M^{i_1}\left(\ker{}^{\psi}D_{\alpha}\right). \tag{2.44}$$

The inclusion (2.44) signifies that there exists a function $g \in \ker^{\psi}D_\alpha$ such that $f = gi_1$. But then $fi_1 = -g$ and therefore $(fi_1) \in \ker^{\psi}D_\alpha$.

Thus we have that $f \in \ker^{\psi}\Pi^+_\alpha$ is equivalent to (2.44). □

**2.19** Following the ideas described in the Introduction, we construct a function theory associated with the Helmholtz operator. This requires a fundamental solution for the Helmholtz operator with $\lambda \in \mathbb{H}(\mathbb{C})$. For the particular case of $\lambda \in \mathbb{C}$ such a fundamental solution is well known: if $\lambda = \nu^2 \in \mathbb{C}$, a fundamental solution $\theta_\nu$ of the Helmholtz operator $\Delta_{\nu^2} := \Delta + \nu^2 I$ is given by the formula

$$\theta_\nu(x) := -(4\pi|x|)^{-1} \cdot e^{-i\nu\cdot|x|}, \quad x \in \mathbf{R}^3\backslash\{0\}; \tag{2.45}$$

see, e.g., [130]. In particular, for $\nu = 0$ we get a fundamental solution of the Laplace operator:

$$\theta(x) := \theta_0(x) = -\frac{1}{4\pi|x|}.$$

**2.20** Now let $\lambda = \alpha^2 \notin \mathfrak{S}, \vec{\alpha}^2 \neq 0$. Write $\gamma := +\sqrt{\vec{\alpha}^2} \in \mathbb{C}$; $\xi_\pm := \alpha_0 \pm \gamma$. Let us introduce the operators

$$P^\pm := \frac{1}{2\gamma} M^{(\gamma \pm \vec{\alpha})},$$

which are mutually complementary projectors (Proposition 2.11). Then

$$\alpha^2 = \lambda = P^+ \left[\xi_+^2\right] + P^- \left[\xi_-^2\right].$$

Therefore

$$\Delta + M^\lambda = P^+ \left(\Delta + \xi_+^2 I\right) + P^- \left(\Delta + \xi_-^2 I\right). \tag{2.46}$$

Formula (2.46) gives a simple way to construct a fundamental solution of the operator $\Delta + M^\lambda$ in the case under consideration. Namely, if for complex numbers $\xi_\pm$, $\theta_{\xi_\pm}$ denotes the functions from (2.45) then the function

$$\theta_\alpha := P^+\left[\theta_{\xi_+}\right] + P^-\left[\theta_{\xi_-}\right]$$

is a fundamental solution of the operator $\Delta + M^{\alpha^2}$. In fact,

$$\left(\Delta + M^{\alpha^2}\right)[\theta_\alpha] = P^+\left(\Delta + \xi_+^2 I\right)\left[\theta_{\xi_+}\right] + P^-\left(\Delta + \xi_-^2 I\right)\left[\theta_{\xi_-}\right] = \delta,$$

where $\delta$ is the delta function.

**2.21** We now treat the case

$$\lambda \notin \mathfrak{S},\ \vec{\alpha}^2 = 0.$$

Let us verify that

$$\begin{aligned} \theta_\alpha(x) : &= \theta_{\alpha_0}(x) + \frac{\partial}{\partial \alpha_0}\left[\theta_{\alpha_0}(x)\right]\vec{\alpha} = \\ &= \theta_{\alpha_0}(x) - i\theta_{\alpha_0}(x)\cdot |x|\vec{\alpha} \end{aligned}$$

is a fundamental solution of the operator $\Delta + M^{\alpha^2}$. Note that $\alpha^2 = \alpha_0^2 + 2\alpha_0\vec{\alpha}$.

Then

$$\left(\Delta + M^{\alpha^2}\right)[\theta_\alpha](x) =$$

$$= \left(\Delta + \alpha_0^2 I + 2\alpha_0 M^{\vec{\alpha}}\right)\left[\theta_{\alpha_0}(x) + \frac{\partial}{\partial \alpha_0}\left(\theta_{\alpha_0}(x)\right)\vec{\alpha}\right] =$$

$$= \delta(x) + 2\alpha_0\theta_{\alpha_0}(x)\vec{\alpha} + \frac{\partial}{\partial \alpha_0}\left(\Delta\left[\theta_{\alpha_0}\right](x)\right)\vec{\alpha} +$$

$$+ \frac{\partial}{\partial \alpha_0}\left(\alpha_0^2\theta_{\alpha_0}(x)\right)\vec{\alpha} - 2\alpha_0\theta_{\alpha_0}(x)\vec{\alpha} =$$

$$= \delta(x) + \frac{\partial}{\partial \alpha_0}\left(\delta(x)\right)\vec{\alpha} =$$

$$= \delta(x).$$

**2.22** To complete the proof we have to consider $\lambda \in \mathfrak{S}$. Noting that $\alpha_0 \neq 0$ and using Proposition 2.11 with

$$P^+ := \frac{1}{2\alpha_0} M^{\alpha}, \; P^- := \frac{1}{2\alpha_0} M^{\bar{\alpha}},$$

we can write

$$\alpha^2 = 2\alpha_0\alpha = P^+\left[4\alpha_0^2\right].$$

This implies the following representation of $\Delta_\lambda$ :

$$\Delta_\lambda = \Delta + M^{\alpha^2} = P^+ \cdot \Delta_{4\alpha_0^2} + P^- \cdot \Delta, \tag{2.47}$$

which gives immediately

$$\theta_\alpha = P^+\left[\theta_{2\alpha_0}\right] + P^-\left[\theta_0\right].$$

We arrive at the theorem.

**2.23 Theorem** (Fundamental solution of the Helmholtz equation with a quaternionic wave number) *Let $\lambda \in \mathbb{H}(\mathbb{C})$. Then a fundamental solution $\theta_\alpha$ for the operator $\Delta_{\alpha^2} = \Delta + M^{\alpha^2}$ is given by the following formulas: for $x \in \mathbf{R}^3\backslash\{0\}$,*

1. *if $\alpha^2 \notin \mathfrak{S}$, $\vec{\alpha}^2 \neq 0$, then*

$$\theta_\alpha(x) = \frac{1}{2\gamma}\Big(\theta_{\xi_+}(x)(\gamma + \vec{\alpha}) + \theta_{\xi_-}(x)(\gamma - \vec{\alpha})\Big);$$

2. *if $\alpha^2 \notin \mathfrak{S}, \vec{\alpha}^2 = 0$, then*

$$\theta_\alpha(x) = \theta_{\alpha_0}(x) - i\theta_{\alpha_0}(x) \cdot |x| \cdot \vec{\alpha};$$

3. *if $\alpha^2 \in \mathfrak{S}$, then*

$$\theta_\alpha(x) = \frac{1}{2\alpha_0}\left(\theta_{2\alpha_0}(x) \cdot \alpha + \theta(x) \cdot \bar{\alpha}\right);$$

*here $\theta_{\xi_\pm}, \theta_{2\alpha_0}, \theta_{\alpha_0}, \theta$ are defined in Subsection 2.19.*

# 3 Notion of $\alpha$-holomorphic function

**3.1** Let $\psi$ be a structural set, $\alpha \in \mathbb{H}(\mathbb{C})$, and $\Omega$ a domain in $\mathbb{R}^3$. For the operator ${}^{\psi}D_{\alpha}$ introduced in Subsection 2.6 we define a set

$$ {}^{\psi}\mathfrak{M}_{\alpha} := {}^{\psi}\mathfrak{M}_{\alpha}(\Omega; \mathbb{H}(\mathbb{C})) := \ker {}^{\psi}D_{\alpha}. \tag{3.1} $$

Its elements sometimes will be referred to as left $(\psi, \alpha)$-hyperholomorphic functions. For the standard structural set $\psi_{st} := \{i_1, i_2, i_3\}$ we use the notation $D := {}^{\psi_{st}}D$; $D_{\alpha} := {}^{\psi_{st}}D_{\alpha}$; $\mathfrak{M}_{\alpha} := {}^{\psi_{st}}\mathfrak{M}_{\alpha}$. For brevity we shall write simply "holomorphic", omitting "left" and "$(\psi, \alpha)$-" if misunderstanding cannot arise, specifying it only if necessary. For instance, $\alpha$-holomorphic function means a function from $\mathfrak{M}_{\alpha}$. Each of the sets (3.1) will be called a class of quaternionic holomorphy.

Of course, we can obtain "symmetric" definitions exchanging the left Cauchy–Riemann operator ${}^{\psi}D_{\alpha}$ for the right one, ${}_{\alpha}D^{\psi} := {}^{\alpha}M - D^{\psi}$. It is obvious that the operator $D$ coincides with the Moisil–Theodoresco operator (2.3). Although we are going to pay most attention to the case of $\alpha \neq 0$, the essential part of the results is true for $\alpha = 0$ including the case of quaternionic holomorphic functions in the sense of Moisil–Theodoresco.

The equality (2.18) means that each quaternionic holomorphic function in $C^2(\Omega, \mathbb{H}(\mathbb{C}))$ is metaharmonic. Below we shall prove that in fact each quaternionic holomorphic function is of class $C^2$ and, hence, is metaharmonic.

**3.2** If $\alpha$ belongs to $\mathbb{H}(\mathbb{R})$, then it is obvious that

$$ {}^{\psi}\mathfrak{M}_{\alpha}(\Omega; \mathbb{H}(\mathbb{C})) = {}^{\psi}\mathfrak{M}_{\alpha}(\Omega; \mathbb{H}(\mathbb{R})) \pm i\,{}^{\psi}\mathfrak{M}_{\alpha}(\Omega; \mathbb{H}(\mathbb{R})). $$

Thus for $\alpha \in \mathbb{H}(\mathbb{R})$ it is enough to consider $\mathbb{H}(\mathbb{R})$-valued hyperholomorphic functions. This follows from the fact that the operator ${}^{\psi}D$ maps $\mathbb{H}(\mathbb{R})$-valued functions onto $\mathbb{H}(\mathbb{R})$-valued ones.

The following theorem is a reformulation of Theorem 2.13 in new terms.

**3.3 Theorem** (Metaharmonic and $\alpha$-holomorphic functions) *Let $\lambda \neq 0$, and let $\mathcal{H}_\lambda$ be the space of metaharmonic functions. Then*

*for $\lambda \notin \mathfrak{S}$,*

$$\mathcal{H}_\lambda = {}^{\psi}\mathfrak{M}_\alpha \bigoplus {}^{\bar{\psi}}\mathfrak{M}_{-\alpha},$$

*for $\lambda \in \mathfrak{S}$,*

$$\mathcal{H}_\lambda = {}^{\psi}\mathfrak{M}_{2\alpha_0} \cdot \alpha \bigoplus {}^{\bar{\psi}}\mathfrak{M}_{-2\alpha_0} \cdot \alpha \bigoplus \mathcal{H} \cdot \bar{\alpha}.$$

Theorem 2.18 shows that there exists a paradoxical relation between **harmonic** and $(\psi, \alpha)$–holomorphic functions even if $\alpha \neq 0$:

**3.4 Theorem** (Harmonic and $(\psi, \alpha)$-holomorphic functions) *Let $\alpha \in \mathfrak{S}$, $\alpha_0 = 0$, and $\alpha_1 \neq 0$. Then*

$$\mathcal{H} = {}^{\psi}\mathfrak{M}_{-\alpha} \bigoplus {}^{\bar{\psi}}\mathfrak{M}_\alpha \cdot i_1.$$

**3.5** We consider now the connections between the classes of hyperholomorphy and structural sets generating them.

Let $\psi$ be a structural set, and let $h \in \mathbb{H}(\mathbb{R})$, $|h| = 1$. Since the operator of multiplication by $h$ (whether from the left-hand side or from the right-hand side) is an orthogonal transformation of $\mathbb{R}^4$, the sets $\varphi := h \cdot \psi$ and $\tilde{\varphi} := \psi \cdot h$ are structural also. In this case we shall call $\varphi$ and $\psi$ (respectively $\tilde{\varphi}$ and $\psi$) left- (respectively right-) equivalent. We have

$$\begin{aligned} {}^{h\psi}D &= \sum_{k=1}^{3} h \cdot \psi^k \cdot \partial_k = h \cdot {}^{\psi}D, \\ D^{\psi h} &= M^h \cdot D^{\psi}. \end{aligned}$$

Thus

$$^{\psi}\mathfrak{M} = {}^{h\psi}\mathfrak{M}; \quad \mathfrak{M}^{\psi} = \mathfrak{M}^{\psi h}.$$

In exactly the same way, from

$$D^{h\psi} = D^{\psi} \cdot M^{h}; \quad {}^{\psi h}D = {}^{\psi}D \cdot {}^{h}M$$

it follows that

$$\mathfrak{M}^{\psi} = \mathfrak{M}^{h\psi} \cdot h; \quad {}^{\psi}\mathfrak{M} = h \cdot {}^{\psi h}\mathfrak{M}.$$

Thus a class of hyperholomorphy $\left({}^{\psi}\mathfrak{M} \quad \text{or} \quad \mathfrak{M}^{\psi}\right)$ does not change under:

1. multiplication of left (respectively right-) $\psi$-holomorphic functions from the right-hand side (respectively from the left-hand side) by constants from $\mathbb{H}(\mathbb{C})$,

2. replacing a structural set by an equivalent one,

3. simultaneous multiplication "from the wrong side" by $h \in \mathbb{S}^3 \subset \mathbb{H}$ of all functions of a class and of a structural set.

We note also that from (2.16) we have

$${}^{\psi}\mathfrak{M} = Z_{\mathbb{H}}\left(\mathfrak{M}^{\bar{\psi}}\right); \quad \mathfrak{M}^{\psi} = Z_{\mathbb{H}}\left({}^{\bar{\psi}}\mathfrak{M}\right).$$

Now let $a \in \mathbb{H}(\mathbb{R})\backslash\{0\}$. Since all classes of $\psi$-hyperholomorphy are $\mathbb{R}$-linear, for $h := \frac{1}{|a|}$ we have the following: if $f \in {}^{\psi}\mathfrak{M}$, then $af \notin {}^{\psi}\mathfrak{M}$, in general, but $af = |a| \cdot hf \in {}^{\psi\bar{h}}\mathfrak{M}$. So multiplying a left $\psi$-holomorphic function by a quaternion from the left-hand side, we get a left-holomorphic function but generally from another class. The same holds in the right case.

**3.6** We shall describe now how an affine change of variables (with an orthogonal matrix) affects the operator ${}^{\psi}D$ with fixed $\psi$. This will allow us to point out the connection between an arbitrary $\psi$-hyperholomorphy and standard hyperholomorphy.

Let $\Omega \subset \mathbb{R}^3_x$, and let $x = \gamma(y) := ay + b$ where $a = (a_{kq}) \in O_3(\mathbb{R})$, $b \in \mathbb{R}^3$, $y \in \gamma^{-1}(\Omega) =: \Xi$.

Let us denote by $W_\gamma$ the operator of change of variable: $W_\gamma[f](y) = g = (f \circ \gamma)(y)$. As one can easily see we have

$$\left(W_\gamma \cdot {}^\psi D_x\right)[f](y) = \sum_{q=1}^{3} \left(\sum_{k=1}^{3} \psi^k \cdot a_{qk}\right) \cdot \partial_{y^q} f(y).$$

It is easily seen that the set $\varphi := \{\varphi^1, \varphi^2, \varphi^3\}$, with

$$\varphi^q := \sum_{k=1}^{3} \psi^k \cdot a_{qk},$$

is structural. Consequently, the operator ${}^\psi D_x$ is converted into ${}^\varphi D_y$:

$$W_\gamma \cdot {}^\psi D_x = {}^\varphi D_y \cdot W_\gamma,$$

and

$$f \in {}^\psi\mathfrak{M}(\Omega) \iff f \circ \gamma \in {}^\psi\mathfrak{M}(\Xi).$$

**3.7** As was established above, equivalent structural sets define the same class of hyperholomorphy. Now we show that among all equivalent structural sets there is one which consists of purely imaginary quaternions. In fact, for any fixed structural set $\psi$ there are exactly two quaternions $\pm\psi^0$ such that sets of vectors $\{\psi^0, \psi^1, \psi^2, \psi^3\}$ form orthonormalized bases in $\mathbb{R}^4_y$. Let $\psi^0$ be chosen in such a way that the basis $\{\psi^0, \psi^1, \psi^2, \psi^3\}$ is cooriented with the canonical basis in $\mathbb{R}^4_y$, and denote $h := \bar{\psi}^0$. Then $\varphi^0 := h\psi^0 = i_0$, whence the set $\varphi := h \cdot \psi$ which is left equivalent to $\psi$ lies on the hyperplane $\tilde{R}^3_y := \{(0, y^1, y^2, y^3) \,|\, (y^1, y^2, y^3) \in \mathbb{R}^3\}$.

Analogously if $\lambda := \psi h$ then the $\lambda$ which is right equivalent to $\psi$ lies in $\tilde{R}^3_y$.

It then follows that a class of $\psi$-hyperholomorphy is defined in fact by a certain orthonormal basis in $\mathbb{R}^3$ (but not by three orthonormal vectors in $\mathbb{R}^4 \approx \mathbb{H}(\mathbb{R})$).

Let $\psi = \{\psi^1, \psi^2, \psi^3\}$ be a structural set with all $\psi^k$ purely imaginary. Let

$$a := \left(\psi_j^k\right)_{k,j=1}^3 .$$

Then $a \in O_3^+(\mathbb{R})$. Consider the change of variable $x = a \cdot y$. Formulas from Subsection 3.6 give

$$\varphi^q = \sum_{k=1}^3 \psi^k \cdot \psi_q^k = i_q, \quad q \in \mathbb{N}_3.$$

Summarizing we thus have by a linear change of variables $\gamma$ such that

$$f \in {}^{\psi}\mathfrak{M}(\Omega) \Longleftrightarrow f \circ \gamma \in \mathfrak{M}\left(\gamma^{-1}(\Omega)\right).$$

Hence each $\psi$-holomorphic function is standard-holomorphic but in another domain. This shows that we have at least two ways of developing the theory of quaternionic holomorphic functions: the first is to develop it for an arbitrary $\psi$, and the second consists of developing the case of $\psi = \psi_{st} = (i_1, i_2, i_3)$ reducing then the general case to this particular one and seeing what we obtain after the necessary transformations.

We will supplement this by some concrete examples later.

**3.8** The contents of Subsections 3.5–3.7 can be easily applied to the operator ${}^{\psi}D_\alpha$. For instance, under the notation of Subsection 3.6 we can write

$$f \in {}^{\psi}\mathfrak{M}_\alpha(\Omega) \Longleftrightarrow f \circ \gamma \in {}^{\varphi}\mathfrak{M}_\alpha(\Xi).$$

**3.9** It is known that if $\lambda \in \mathbb{C}$ (and if, in particular, $\lambda \in \mathbb{R}$) then $\mathbb{C}$-valued metaharmonic functions exist. Let $f_0$ be one of them and let

$$g(x) := -{}^{\psi}D_{-\alpha}[f_0](x) = \alpha f_0 - \sum_{k=1}^3 \psi^k \cdot \partial_k f_0 =$$
$$= -{}_{-\alpha}D^{\psi}[f_0](x).$$

Then ${}^{\psi}D_\alpha[g] = {}^{\psi}D_\alpha \cdot \left(-{}^{\psi}D_{-\alpha}\right)[f_0] = \Delta_\lambda[f_0] = 0$. By analogy, ${}_{\alpha}D^{\psi}[g] = 0$.

Thus it has been shown that for $\lambda \in \mathbb{C}$, classes of $(\psi,\alpha)$-holomorphic functions are quite large and that

$$ {}^{\psi}\mathfrak{M}_{\alpha} \bigcap {}_{\alpha}\mathfrak{M}^{\psi} \neq \emptyset. $$

From Theorem 3.3 it is clear "how large" the set of all metaharmonic functions is, for $\lambda \in \mathfrak{S}$. If $\lambda \notin \mathfrak{S}$ then the equality (2.46) demonstrates that corresponding metaharmonic functions exist, and the set ${}^{\psi}\mathfrak{M}_{\alpha}$ is "large enough".

**3.10** We can now obtain different formulations for the analogue of the Cauchy–Riemann conditions for a function $f$ to be left $(\psi,\alpha)$-holomorphic, that is, different correlations for components of $f$.

First of all, omitting the multiplicative quaternionic structure in $\mathbb{C}^4$, we come to the following matrix partial differential operator acting on $C^1(\Omega;\mathbb{C}^4)$:

$$ {}^{\psi}\bar{D}_{\alpha} := $$

$$ \begin{pmatrix} \alpha_0; & -\sum_{k=1}^{3}\psi_1^k\partial_k - \alpha_1; & -\sum_{k=1}^{3}\psi_2^k\partial_k - \alpha_2; & -\sum_{k=1}^{3}\psi_3^k\partial_k - \alpha_3 \\ \sum_{k=1}^{3}\psi_1^k\partial_k + \alpha_1; & \alpha_0; & -\sum_{k=1}^{3}\psi_3^k\partial_k + \alpha_3; & \sum_{k=1}^{3}\psi_2^k\partial_k - \alpha_2 \\ \sum_{k=1}^{3}\psi_2^k\partial_k + \alpha_2; & \sum_{k=1}^{3}\psi_3^k\partial_k - \alpha_3; & \alpha_0 & -\sum_{k=1}^{3}\psi_1^k\partial_k + \alpha_1 \\ \sum_{k=1}^{3}\psi_3^k\partial_k + \alpha_3; & -\sum_{k=1}^{3}\psi_2^k\partial_k + \alpha_2; & \sum_{k=1}^{3}\psi_1^k\partial_k - \alpha_1; & \alpha_0 \end{pmatrix}. $$

It is clear that ${}^{\psi}\bar{D}_{\alpha}$ is an elliptic operator. Separating components in the equality ${}^{\psi}\bar{D}_{\alpha}[f] = 0$ we get the following analogue of the Cauchy–Riemann conditions:

$$\begin{cases} \alpha_0 f_0 - \sum_{p=1}^{3}\left(\sum_{k=1}^{3}\psi_p^k \cdot \partial_k f_p + \alpha_p f_p\right) = 0, \\ \alpha_0 f_1 + \alpha_1 f_0 - \alpha_2 f_3 + \sum_{k=1}^{3}\left(\psi_1^k \cdot \partial_k f_0 - \psi_3^k \cdot \partial_k f_2 + \psi_2^k \cdot \partial_k f_3\right) = 0, \\ \alpha_2 f_0 + \alpha_0 f_2 - \alpha_3 f_1 + \sum_{k=1}^{3}\left(\psi_2^k \cdot \partial_k f_0 + \psi_3^k \cdot \partial_k f_1 - \psi_1^k \partial_k f_3\right) = 0, \\ \alpha_3 f_0 + \alpha_0 f_3 + \alpha_2 f_1 + \sum_{k=1}^{3}\left(\psi_3^k \cdot \partial_k f_0 - \psi_2^k \cdot \partial_k f_1 + \psi_1^k \cdot \partial_k f_2\right) = 0. \end{cases}$$

**3.11** Let now $f_0$ and $\vec{f}$ be "vector components" of an $\alpha$-holomorphic function; i.e., $\psi = \psi_{st}$. Then

$$D_\alpha\left[f_0 + \vec{f}\right] = -\text{div}\vec{f} - <\vec{f}, \vec{\alpha}> + \alpha_0 f_0 +$$

$$+\text{grad } f_0 + \text{rot}\vec{f} + \left[\vec{f} \times \vec{\alpha}\right] + f_0\vec{\alpha} + \alpha_0\vec{f}.$$

Separating the scalar part and the vector part in the equality $D_\alpha[f] = 0$ we arrive at the vector form of the Cauchy–Riemann conditions:

$$\begin{cases} \alpha_0 f_0 - \text{div}\vec{f} - <\vec{f}, \vec{\alpha}> = 0, \\ \text{grad } f_0 + \text{rot}\vec{f} + \left[\vec{f} \times \vec{\alpha}\right] + f_0\vec{\alpha} + \alpha_0\vec{f} = 0. \end{cases} \tag{3.2}$$

In particular, a vector field $\vec{f}$ is $\alpha$-holomorphic if

$$\begin{cases} \text{div}\vec{f} + <\vec{f}, \vec{\alpha}> = 0, \\ \text{rot}\vec{f} + \left[\vec{f} \times \vec{\alpha}\right] + \alpha_0\vec{f} = 0. \end{cases}$$

When $\alpha = 0$ we obtain a system defining solenoidal and irrotational vector fields. When in (3.2) $\alpha_0 = 0$, $\alpha_k \in \mathbb{R}$, $k \in \mathbb{N}_3$, the system (3.2) defines the so-called generalized holomorphic vectors in the sense of A. V. Bitsadze which were studied in [94, 77, 134], and in many other works. It should be noticed that the investigation of the generalized holomorphic vectors was carried out in matrix language. The theories associated with the operators $D_{\alpha_0}$ and $D_{\vec{\alpha}}$, $\alpha_0 \in \mathbb{R}$; $\vec{\alpha} \in \mathbb{R}^3$, have been developed along similar lines (see [45, 46]; [94, 77, 134]); similar results were obtained but the two theories had no intersection.

As is seen from (3.2) both cases are included in our considerations. Moreover, in Subsection 3.19 we give a simple connection between the two theories which annihilates the difference between them. In Section 9 we show how both cases are related to the Maxwell equations.

**3.12** The next section is devoted to the main integral representations of the holomorphic function theory. This requires some useful formulas involving differential form techniques. Let $dx = dx_1 \wedge dx_2 \wedge dx_3$ be the differential volume form in $\mathbb{R}^3$. Denote, as usual, by $d\hat{x}^k$ the differential form $dx$ with a factor $dx^k$ omitted; i.e., $d\hat{x}^1 := dx^2 \wedge dx^3$, etc. Introduce

$$\sigma_{\psi,x} := \sum_{k=1}^{3}(-1)^{k-1}\psi^k \cdot d\hat{x}^k.$$

Sometimes we shall omit the index "$x$" in $\sigma_{\psi,x}$. Let $ds$ be the surface area element in $\mathbb{R}^3$. Then $|\sigma_{\psi,x}| = ds$, and if $\Gamma$ is a smooth surface, then

$$\sigma_{\psi,x} = \sum_{k=1}^{3}\psi^k \cdot n_k(x)ds = n_\psi(x)ds$$

where $\vec{n} = (n_1, n_2, n_3)$ is the unit vector of the outward normal on $\Gamma$ at the point $x \in \Gamma$, $n_\psi := \sum_{k=1}^{3}\psi^k n_k$.

It is natural to call $n_\psi$ the quaternionic representation of the unit normal, as well as to call $\sigma_{\psi,x}$ the quaternionic surface area element in $\mathbb{R}^3$.

**3.13** Let $\{f, g\} \subset C^1\left(\bar{\Omega}, \mathbb{H}(\mathbb{C})\right)$. Direct computation leads to the equality

$$\begin{aligned} d(g \cdot \sigma_\psi \cdot f) &= dg \wedge \sigma_\psi f - g\sigma_\psi \wedge df = \\ &= \left(D^\psi[g] \cdot f + g \cdot {}^\psi D[f]\right) dx \end{aligned}$$

where $d$ denotes the operator of exterior differentiation of a differential form. Taking into account how ${}^{\psi}D$ and ${}^{\psi}D_{\alpha}$ are connected we obtain

$$d(g\cdot\sigma_{\psi}\cdot f)=\begin{cases}\left({}_{\alpha}D^{\psi}[g]\cdot f+g\cdot{}^{\psi}D_{\alpha}[f]-(\alpha g f+g f\alpha)\right)dx,\\ \\ \left({}_{\alpha}D^{\psi}[g]\cdot f-g\cdot{}^{\bar{\psi}}D_{\alpha}[f]-(\alpha g f-g f\alpha)\right)dx,\\ \\ \left(g\cdot{}^{\psi}D_{\alpha}[f]-{}_{\alpha}D^{\bar{\psi}}[g]\cdot f-(g f\alpha-\alpha g f)\right)dx.\end{cases}$$

Note that for $\alpha\in\mathbb{C}$ all formulas become simpler:

$$d(g\cdot\sigma_{\psi}\cdot f)=\begin{cases}\left({}_{\alpha}D^{\psi}[g]\cdot f+g\cdot{}^{\psi}D_{\alpha}[f]-2\alpha g f\right)dx,\\ \\ \left({}_{\alpha}D^{\psi}[g]\cdot f-g\cdot{}^{\bar{\psi}}D_{\alpha}[f]\right)dx,\\ \\ \left(g\cdot{}^{\psi}D_{\alpha}[f]-{}_{\alpha}D^{\bar{\psi}}[g]\cdot f\right)dx.\end{cases}$$

**3.14 Proposition** (Definition of hyperholomorphy in terms of differential forms) *Let* $\{f,g\}\subset C^{1}\left(\bar{\Omega};\mathbb{H}(\mathbb{C})\right)$. *Then*

1. $f\in{}^{\psi}\mathfrak{M}_{\alpha}\Longleftrightarrow d(\sigma_{\psi}\cdot f)=-f\alpha dx$,

2. $g\in{}_{\alpha}\mathfrak{M}^{\psi}\Longleftrightarrow d(g\cdot\sigma_{\psi})=-\alpha g dx$,

3. $f\in{}^{\psi}\mathfrak{M}_{\alpha}$ *and* $g\in{}_{\alpha}\mathfrak{M}^{\psi}\Longleftrightarrow d(g\cdot\sigma_{\psi}\cdot f)=-(\alpha g f+f g\alpha)\,dx$.

PROOF. This is a direct corollary of Subsection 3.13. □

**3.15** We now construct an $\alpha$-holomorphic function which plays a remarkable role in quaternionic function theory, namely, the analogue of the Cauchy kernel for one-dimensional complex analysis. Denote by $\mathcal{K}_{\psi,\alpha}$ the function defined by the formula

$$\mathcal{K}_{\psi,\alpha} = \begin{cases} -{}^{\psi}D_{-\alpha}\left[\theta_\alpha\right], & \alpha \in G\mathbb{H}(\mathbb{C}) \quad \text{or} \\ & \alpha \in \mathfrak{S} \quad \text{but} \quad \alpha_0 \neq 0; \\ -{}^{\psi}D_{-\alpha}[\theta], & \alpha \in \mathfrak{S} \quad \text{and} \\ & \alpha_0 = 0 \Longleftrightarrow \vec{\alpha}^2 = 0, \\ -{}^{\psi}D[\theta], & \alpha = 0. \end{cases} \tag{3.3}$$

Theorem 2.23 implies immediately that $\mathcal{K}_{\psi,\alpha}$ is a fundamental solution for the operator ${}^{\psi}D_\alpha$ for any $\alpha \in \mathbb{H}(\mathbb{C})$. Thus the following theorem holds:

**3.16 Theorem** (Fundamental solution for the operator ${}^{\psi}D_\alpha$, or an analogue of the Cauchy kernel for $\alpha$-holomorphic function theory) *Let $\alpha \in \mathbb{H}(\mathbb{C})$. Then a fundamental solution for the operator ${}^{\psi}D_\alpha$ is given by the formulas:*
*for $x \in \mathbb{R}^3 \setminus \{0\}$,*

1. *if $\alpha = \alpha_0 \in \mathbb{C}$, then*

$$\mathcal{K}_{\psi,\alpha}(x) = \theta_\alpha(x)\left(\alpha + \frac{x_\psi}{|x|^2} + i\alpha\frac{x_\psi}{|x|}\right), \tag{3.4}$$

*where $x_\psi := \sum_{k=1}^{3} \psi^k x_k$;*

2. *if $\alpha \notin \mathfrak{S}$, $\vec{\alpha}^2 \neq 0$, then*

$$\mathcal{K}_{\psi,\alpha}(x) = P^+\left[\mathcal{K}_{\psi,\xi_+}\right](x) + P^-\left[\mathcal{K}_{\psi,\xi_-}\right](x), \tag{3.5}$$

*where $\mathcal{K}_{\psi,\xi_\pm}$ are the fundamental solutions for the operators ${}^{\psi}D_{\xi_\pm}$ with the complex parameters $\xi_\pm$, defined by (3.4) (see the definition of $P^\pm$ in Subsection 2.20);*

3. *if $\alpha \notin \mathfrak{S}$, $\vec{\alpha}^2 = 0$, then*

$$\mathcal{K}_{\psi,\alpha} = \mathcal{K}_{\psi,\alpha_0} + \frac{\partial}{\partial \alpha_0}\left[\mathcal{K}_{\psi,\alpha_0}\right]\vec{\alpha};$$

4. *if $\alpha \in \mathfrak{S}$, $\alpha_0 \neq 0$, then*

$$\mathcal{K}_{\psi,\alpha} = P^+\left[\mathcal{K}_{\psi,2\alpha_0}\right] + P^-\left[\mathcal{K}_{\psi,0}\right];$$

*5. if* $\alpha \in \mathfrak{S}$, $\alpha_0 = 0$, *then*

$$\mathcal{K}_{\psi,\alpha} = \mathcal{K}_{\psi,0} + \theta \cdot \alpha.$$

PROOF. Formula (3.4) is obtained from the definition (3.3) by a simple straightforward calculation.

To obtain (3.5) we first note that

$$ {}^{\psi}D_{-\alpha} \cdot P^{\pm} = P^{\pm} \cdot {}^{\psi}D_{-\alpha} = P^{\pm} \cdot {}^{\psi}D_{-\xi_{\pm}}.$$

Then (using Subsection 2.20)

$$-{}^{\psi}D_{-\alpha}\left[\theta_{\alpha}\right] = -{}^{\psi}D_{-\alpha}\left[P^{+}\left[\theta_{\xi_+}\right] + P^{-}\left[\theta_{\xi_-}\right]\right] =$$

$$= P^{+}\left[-{}^{\psi}D_{-\xi_+}\left[\theta_{\xi_+}\right]\right] + P^{-}\left[-{}^{\psi}D_{-\xi_-}\left[\theta_{\xi_-}\right]\right] =$$

$$= P^{+}\left[\mathcal{K}_{\psi,\xi_+}\right] + P^{-}\left[\mathcal{K}_{\psi,\xi_-}\right].$$

When $\alpha \notin \mathfrak{S}$, $\vec{\alpha}^2 = 0$ Subsection 2.21 will help us. We have:

$$-{}^{\psi}D_{-\alpha}\left[\theta_{\alpha}\right] = -{}^{\psi}D_{-\alpha}\left[\theta_{\alpha_0} + \frac{\partial}{\partial\alpha_0}\left[\theta_{\alpha_0}\right]\vec{\alpha}\right] =$$

$$= -{}^{\psi}D_{-\alpha_0}\left[\theta_{\alpha_0}\right] + \theta_{\alpha_0}\vec{\alpha} - {}^{\psi}D_{-\alpha_0}\left[\frac{\partial}{\partial_{\alpha_0}}\left[\theta_{\alpha_0}\right]\right]\vec{\alpha}.$$

From the evident equality

$$\alpha_0\frac{\partial}{\partial\alpha_0}\left[\theta_{\alpha_0}\right] = \frac{\partial}{\partial\alpha_0}\left[\alpha_0\theta_{\alpha_0}\right] - \theta_{\alpha_0}$$

we obtain

$$-{}^{\psi}D_{-\alpha}\left[\theta_{\alpha}\right] = \mathcal{K}_{\psi,\alpha_0} + \theta_{\alpha_0}\vec{\alpha} + \frac{\partial}{\partial\alpha_0}\left[-{}^{\psi}D_{-\alpha_0}\left[\theta_{\alpha_0}\right]\right]\vec{\alpha} - -\theta_{\alpha_0}\vec{\alpha} =$$

$$= \mathcal{K}_{\psi,\alpha_0} + \frac{\partial}{\partial\alpha_0}\left[\mathcal{K}_{\psi,\alpha_0}\right]\vec{\alpha}.$$

When $\alpha \in \mathfrak{S}$, $\alpha_0 \neq 0$ we have (see Subsection 2.22)

$$\begin{aligned} -{}^{\psi}D_{-\alpha}[\theta_\alpha] &= -{}^{\psi}D_{-\alpha}\left[P^+[\theta_{2\alpha_0}] + P^-[\theta]\right] = \\ &= P^+\left[-{}^{\psi}D_{-2\alpha_0}[\theta_{2\alpha_0}]\right] + P^-\left[-{}^{\psi}D[\theta]\right] = \\ &= P^+[\mathcal{K}_{\psi,2\alpha_0}] + P^-[\mathcal{K}_{\psi,0}]. \end{aligned}$$

For the last case $\alpha \in \mathfrak{S}$, $\alpha_0 = 0$ we have

$$-{}^{\psi}D_{-\alpha}[\theta] = -{}^{\psi}D[\theta] + \theta\alpha = \mathcal{K}_{\psi,0} + \theta\alpha.$$

□

**3.17** The fundamental solution $\mathcal{K}_{\psi,\alpha}$ for the operator ${}^{\psi}D_\alpha$, $\alpha \in \mathbb{H}(\mathbb{C})$ can be constructed in another way. Multiplying ${}^{\psi}D_\alpha$ by appropriate differential operators we reduce the problem to constructing the fundamental solution for a differential operator of a higher order but with complex coefficients. For example, when $\alpha \notin \mathfrak{S}$, $\vec{\alpha}^2 \neq 0$, $\alpha_0 \neq 0$, consider the operator

$$A := {}^{\psi}D_\alpha {}^{\psi}D_{\bar{\alpha}} {}^{\psi}D_{-\alpha} {}^{\psi}D_{-\bar{\alpha}}.$$

Using the factorization equality (2.18) and the notation (see Subsection 2.20):

$$\xi_\pm := \alpha_0 \pm \gamma, \qquad \gamma := +\sqrt{\vec{\alpha}^2} \in \mathbb{C},$$

the operator $A$ can be transformed to the form

$$A = \left(\Delta + \xi_+^2\right)\left(\Delta + \xi_-^2\right).$$

Then the function

$$\theta_A := (4\alpha_0\gamma)^{-1}\left(\theta_{\xi_-} - \theta_{\xi_+}\right)$$

is a fundamental solution for the operator $A$. In fact,

$$A[\theta_A] = \left(\xi_+^2 - \xi_-^2\right)^{-1} \left(\left(\Delta + \xi_+^2\right)\left(\Delta + \xi_-^2\right)[\theta_{\xi_-}] - \right.$$

$$\left. - \left(\Delta + \xi_-^2\right)\left(\Delta + \xi_+^2\right)[\theta_{\xi_+}]\right) =$$

$$= \left(\xi_+^2 - \xi_-^2\right)^{-1} \left(\Delta[\delta] + \xi_+^2\delta - \Delta[\delta] - \xi_-^2\delta\right) = \delta.$$

Consequently, the fundamental solution for the operator ${}^{\psi}D_{\alpha}$ can be found by the formula

$$\mathcal{K}_{\psi,\alpha} = (4\alpha_0\gamma)^{-1}\,{}^{\psi}D_{\bar{\alpha}}\,{}^{\psi}D_{-\alpha}\,{}^{\psi}D_{-\bar{\alpha}}\left[\theta_{\xi_-} - \theta_{\xi_+}\right].$$

This method will give us later a factorization of the analogue of the Cauchy type operator (see Theorem 4.18 below).

**3.18** Let us point out that in the case $\alpha \in \mathbb{H}(\mathbb{C})$, $\vec{\alpha}^2 \neq 0$ we have a very useful representation of the operator ${}^{\psi}D_{\alpha}$ through the operators ${}^{\psi}D_{\xi_\pm}$ with the complex parameters $\xi_\pm$:

$$^{\psi}D_{\alpha} = P^+ \cdot {}^{\psi}D_{\xi_+} + P^- \cdot {}^{\psi}D_{\xi_-}. \tag{3.6}$$

Note that when $\alpha \in \mathfrak{S}$, $\vec{\alpha}^2 \neq 0 \iff \alpha_0 \neq 0, \xi_+ = 2\alpha_0, \xi_- = 0$ and (3.6) becomes

$$^{\psi}D_{\alpha} = P^+ \cdot {}^{\psi}D_{2\alpha_0} + P^- \cdot {}^{\psi}D. \tag{3.7}$$

**3.19** Let us distinguish the special case $\alpha_0 = 0$, $\alpha_k \in \mathbb{R}$, $k \in \mathbb{N}_3$. In other words, the case when the equation

$$D_{\alpha}[f] = 0$$

defines the generalized holomorphic vectors in the sense of Bitsadze (see Subsection 3.11). Then the equality (3.6) becomes

$$D_{\vec{\alpha}} = P^+ D_\gamma + P^- D_{-\gamma}. \tag{3.8}$$

An inverse correlation is also true:

$$D_\gamma = P^+ D_{\vec{\alpha}} + P^- D_{-\vec{\alpha}}. \tag{3.9}$$

Note that equalities (3.8), (3.9) establish a connection between the two factorizations (2.18) of the Helmholtz operator $\Delta + \gamma^2 I$, $\gamma \in \mathbb{C}$, $\gamma^2 = \vec{\alpha}^2$. Further, from (3.8), (3.9) it follows that

$$\ker D_{\vec{\alpha}} = P^+ (\ker D_\gamma) \bigoplus P^- (\ker D_{-\gamma}), \tag{3.10}$$

$$\ker D_\gamma = P^+ (\ker D_{\vec{\alpha}}) \bigoplus P^- (\ker D_{-\vec{\alpha}}). \tag{3.11}$$

For example, to prove (3.10) we have

$$D_{\vec{\alpha}}[f] = 0 \Longleftrightarrow P^+ D_\gamma[f] + P^- D_{-\gamma}[f] = 0 \Longleftrightarrow$$

$$\Longleftrightarrow P^+ D_\gamma[f] = 0 \quad \text{and} \quad P^- D_{-\gamma}[f] = 0.$$

The operators $P^\pm$ and $D_{\pm\gamma}$ commute. Therefore

$$P^+ D_\gamma[f] = 0 \Longleftrightarrow P^+[f] \in \ker D_\gamma$$

and

$$P^- D_{-\gamma}[f] = 0 \Longleftrightarrow P^-[f] \in \ker D_{-\gamma},$$

which is equivalent to the inclusions

$$P^+[f] \in P^+ (\ker D_\gamma), \quad P^-[f] \in P^- (\ker D_{-\gamma}).$$

Similarly one can prove (3.11). Therefore the projectors $P^{\pm}$ allow using for $D_\gamma$ any solved problem for $D_{\vec{\alpha}}$ and vice versa.

**3.20 Theorem** (Quaternionic Leibnitz differentiation rule) [46, p. 24] *Let* $\{f, g\} \subset C^1(\Omega; \mathbb{H}(\mathbb{C}))$. *Then*

$$D[f \cdot g] = D[f] \cdot g + \bar{f} \cdot D[g] + 2(\mathrm{Sc}(fD))[g], \tag{3.12}$$

*where* $(\mathrm{Sc}(fD))[g] := -\sum_{k=1}^{3} f_k \partial_k g$.

The equality (3.12) can be verified by straightforward calculation. It means, in particular, that the point-wise product of two holomorphic functions is not, in general, holomorphic.

**3.21** It is natural to consider, together with the operator ${}^{\psi}D_\alpha$, $\alpha \in \mathbb{H}(\mathbb{C})$, the operator

$${}^{\psi}_{\alpha}D := {}^{\psi}D + \alpha I, \quad \alpha \in \mathbb{H}(\mathbb{C}).$$

Then the equation

$${}_{\alpha}D[f] = 0 \tag{3.13}$$

is equivalent to the system

$$\alpha_0 f_0 - \mathrm{div}\vec{f} - < \vec{f}, \vec{\alpha} >= 0, \tag{3.14}$$

$$\mathrm{grad} f_0 + \mathrm{rot}\vec{f} - \left[\vec{f} \times \vec{\alpha}\right] + f_0\vec{\alpha} + \alpha_0\vec{f} = 0, \tag{3.15}$$

which differs from the system (3.2) only by the sign of the third term in equation (3.15). But this small difference affects strongly the properties of the system; this can be seen from the following

**3.22 Proposition**

$$\ker_\alpha D = e^{-<\vec{a},\vec{x}>} I(\ker D_{\alpha_0}).$$

PROOF. Let $g \in \ker D_{\alpha_0}$. We consider the function

$$f(x) := e^{-<\vec{a},\vec{x}>} g(x).$$

Using the Leibnitz rule (3.12) we obtain:

$$\begin{aligned} D[f](x) &= D\left[e^{-<\vec{a},\vec{x}>}\right] g(x) + e^{-<\vec{a},\vec{x}>} D[g] = \\ &= -e^{-<\vec{a},\vec{x}>} \vec{\alpha} g(x) - e^{-<\vec{a},\vec{x}>} \alpha_0 g(x) = -\alpha f(x). \end{aligned}$$

By analogy, for an arbitrary function $f \in \ker_\alpha D$ we have

$$g := e^{<\vec{a},\vec{x}>} f \in \ker D_{\alpha_0}.$$

□

**3.23** This proposition means that the correspondence

$$f \in \ker_\alpha D \longleftrightarrow e^{<\vec{a},\vec{x}>} f =: g \in \ker D_{\alpha_0}$$

defines an isomorphism between $\ker_\alpha D$ and $\ker D_{\alpha_0}$. Hence we have a simple way to reduce the theory associated with the operator ${}_\alpha D$ to that of $D_{\alpha_0}$. We would like to mention that a quite similar situation is valid (compare with [11], see also [56]) for Clifford algebra-valued functions.

**3.24 Füter–type operator** Throughout this book we will denote by $t$ the fourth coordinate of the independent variable which corresponds to the time, and by $\partial_t$ the derivative with respect to $t$. Let us introduce the operator

$$\partial_t - aD, \tag{3.16}$$

defined on $\mathbb{H}(\mathbb{C})$-valued functions from $C^1(\Xi)$, $\Xi \subset \mathbb{R}^4$. Here $a = \mathrm{Const} \in \mathbb{C}$.

The operator (3.16) will be called Füter–type operator ([58]). For $a = -1$ it gives the usual Füter operator. Let us emphasize that the operator (3.16) can be of elliptic type as well as hyperbolic, depending on the parameter $a$. The usual Füter operator is elliptic but, e.g., when $a = \pm i$ we obtain hyperbolic operators which play an important role in the theories of the Maxwell and Dirac equations (see Subsections 8.1, 8.3, 12.5).

A well-developed function theory is associated with the Füter operator $\partial_t + D$ (see, e.g., [113, 114]). A corresponding notion of hypercomplex differentiability can be introduced [83] (see also [79]).

We will examine the operator (3.16) more closely in Section 16.

# 4 Integral formulas for $\alpha$-holomorphic functions

**4.1** We will now derive all essential integral formulas of $(\psi, \alpha)$-holomorphic function theory following the lines described in the Introduction. We need, first of all, a quaternionic version of the Stokes formula.

**4.2 Theorem** (Quaternionic Stokes formula) *Let* $\{f, g\} \subset C^1(\bar{\Omega}; \mathbb{H}(\mathbb{C}))$, $\psi$ *a structural set,* $\alpha \in \mathbb{H}(\mathbb{C})$, *and* $\Gamma = \partial\Omega$ *a piecewise smooth surface. Then*

$$\int_\Gamma g \cdot \sigma_\psi \cdot f = \begin{cases} \int_\Omega ({}_\alpha D^\psi[g] \cdot f + g \cdot {}^\psi\mathcal{D}_\alpha[f] - (\alpha g f + g f \alpha))dx, \\ \\ \int_\Omega ({}_\alpha D^\psi[g] \cdot f - g \cdot {}^{\bar{\psi}}\mathcal{D}_\alpha[f] - (\alpha g f - g f \alpha))dx, \\ \\ \int_\Omega (g\, {}^\psi\mathcal{D}_\alpha[f] - {}_\alpha D^{\bar{\psi}}[g] f - (g f \alpha - \alpha g f))dx. \end{cases} \tag{4.1}$$

PROOF. This is a simple consequence of the usual "real" Stokes theorem and of the contents of Subsection 3.13. □

**4.3 Corollary** (Quaternionic Stokes formula with a complex $\alpha$) *Under the conditions of Theorem 4.2, if* $\alpha \in \mathbb{C}$, *then*

$$\int_\Gamma g\cdot\sigma_\psi\cdot f=\begin{cases}\int_\Omega({}_\alpha D^\psi[g]\cdot f+g\cdot{}^\psi\mathcal{D}_\alpha[f]-2\alpha gf)dx,\\ \\ \int_\Omega({}_\alpha D^\psi[g]\cdot f-g\cdot{}^{\bar\psi}\mathcal{D}_\alpha[f])dx,\\ \\ \int_\Omega(g{}^\psi\mathcal{D}_\alpha[f]-{}_\alpha D^{\bar\psi}[g]\cdot f)dx.\end{cases}\tag{4.2}$$

**4.4** REMARK **Using the representation of the quaternionic differential form $\sigma_\psi$ via the normal (see Subsection 3.12) we can rewrite (4.2) in another form. For instance,**

$$\int_\Gamma g\cdot n_\psi\cdot fd\Gamma=\int_\Omega({}_\alpha D^\psi[g]\cdot f+g{}^\psi\mathcal{D}_\alpha[f]-2\alpha gf)dx.$$

**In particular, the simplest version of the quaternionic Stokes formula is as follows:**

$$\int_\Gamma g\cdot\vec{n}\cdot fd\Gamma=\int_\Omega(D_r[g]\cdot f+g\cdot D[f])dx\tag{4.3}$$

**where $D_r:=D^{st}$, $\vec{n}:=\sum_{k=1}^3 n_k i_k$ is a unit outward normal on $\Gamma$, $\{f,g\}\subset C^1(\bar\Omega;\mathbb{H}(\mathbb{C}))$.**

**4.5** **It should be noted then that integrands on the right-hand sides of (4.1) are equal identically; the same holds for (4.2). However, from (4.1) and (4.2) we can draw different conclusions for the theory of $\alpha$-holomorphic functions. For example, the following assertion holds, which may be considered as an analogue of the Cauchy integral theorem in complex analysis.**

**4.6 Theorem** **(Quaternionic Cauchy integral theorem for conjugate classes of $\alpha$-holomorphy)** *If $g\in{}_\alpha\mathfrak{M}^{\bar\psi}(\bar\Omega;\mathbb{H}(\mathbb{C}))$, $f\in{}^\psi\mathfrak{M}_\alpha(\bar\Omega;\mathbb{H}(\mathbb{C}))$, then*

$$\int_\Gamma g\cdot\sigma_\psi\cdot f=\int_\Omega(\alpha gf-gf\alpha)dx.$$

*In particular, for $\alpha\in\mathbb{C}$,*

$$\int_\Gamma g\cdot\sigma_\psi\cdot f=0\tag{4.4}$$

This follows directly from Theorem 4.2 and Corollary 4.3.

In our opinion, the following assertion, also following from (4.1) and (4.2), is a "more authentic" analogue of the complex Cauchy integral theorem.

**4.7 Theorem** (Quaternionic Cauchy integral theorem for a pair of $(\psi, \alpha)$-holomorphic functions) *Let* $g \in_\alpha \mathfrak{M}^\psi(\bar{\Omega}; \mathbb{H}(\mathbb{C}))$, $f \in^\psi \mathfrak{M}_\alpha(\bar{\Omega}; \mathbb{H}(\mathbb{C}))$. *Then*

$$\int_\Gamma g \cdot \sigma_\psi \cdot f = -\int_\Omega (\alpha g f + g f \alpha) dx. \tag{4.5}$$

*In particular, for* $\alpha \in \mathbb{C}$,

$$\int_\Gamma g \cdot \sigma_\psi \cdot f = -2\alpha \int_\Omega g \cdot f \, dx. \tag{4.6}$$

4.8 Remark If $\alpha = 0$, then ${}^\psi\mathfrak{M} = {}^{\bar{\psi}}\mathfrak{M}$, and Theorem 4.6 and Theorem 4.7 give the same result. Since constants (not being zero divisors) are not $(\psi, \alpha)$-holomorphic functions (if $\alpha \neq 0$), the Cauchy integral theorem for one function does not follow immediately from Theorems 4.6 and 4.7.

Nevertheless it follows from the first equality in (4.1) (see also Proposition 3.14):

**4.9 Theorem** (Quaternionic Cauchy integral theorem for an $\alpha$-holomorphic function) *If* $f \in^\psi \mathfrak{M}_\alpha(\bar{\Omega}; \mathbb{H}(\mathbb{C})$ *and* $g \in_\alpha \mathfrak{M}^\psi(\bar{\Omega}; \mathbb{H}(\mathbb{C}))$, *then*

$$\int_\Gamma \sigma_\psi \cdot f = -\int_\Omega f \cdot \alpha dx; \quad \int_\Gamma g \cdot \sigma_\psi = -\alpha \int_\Omega g dx. \tag{4.7}$$

*In particular, for* $f \in \mathfrak{M}_\alpha(\bar{\Omega}; \mathbb{H}(\mathbb{C}))$

$$\int_\Gamma \vec{n} f d\Gamma = -\int_\Omega f \alpha dx. \tag{4.8}$$

**4.10** We are now ready to construct the quaternionic integral operators which generalize the well-known operators from one-dimensional complex analysis: the Cauchy type operator, the $T$-operator, and the operator of singular integration playing the essential role in what follows.

We start with the case $\alpha = \alpha_0 \in \mathbb{C}$ because all further constructions are based upon it. Let ${}^{\psi}T_{\alpha_0}$, ${}^{\psi}K_{\alpha_0}$ and ${}^{\psi}S_{\alpha_0}$ denote the operators acting by the following rules:

$$ {}^{\psi}T_{\alpha_0}[f](x) := \int_{\Omega} \mathcal{K}_{\psi,\alpha_0}(x-y)f(y)dy, \quad x \in \mathbb{R}^3, \tag{4.9}$$

$$ {}^{\psi}K_{\alpha_0}[f](x) := -\int_{\Gamma} \mathcal{K}_{\psi,\alpha_0}(x-y)n_{\psi}(y)f(y)d\Gamma_y, \ x \in \mathbb{R}^3\backslash\Gamma, \tag{4.10}$$

$$ {}^{\psi}S_{\alpha_0}[f](t) := -2\int_{\Gamma} \mathcal{K}_{\psi,\alpha_0}(t-\tau)n_{\psi}(\tau)f(\tau)d\Gamma_{\tau}, \ \tau \in \Gamma, \tag{4.11}$$

where the Cauchy kernel $\mathcal{K}_{\psi,\alpha_0}$ is defined by the formula (3.4); the integral in (4.11) exists in the sense of the Cauchy principal value.

As usual, the operator of singular integration with the Cauchy kernel generates two other important operators:

$$ {}^{\psi}P_{\alpha_0} := \frac{1}{2}(I + {}^{\psi}S_{\alpha_0}); \quad {}^{\psi}Q_{\alpha_0} := I - {}^{\psi}P_{\alpha_0} = \frac{1}{2}(I - {}^{\psi}S_{\alpha_0}). \tag{4.12}$$

As above, we write $T_{\alpha_0}, K_{\alpha_0}, S_{\alpha_0}, P_{\alpha_0}, Q_{\alpha_0}$ for $\psi = \psi_{st} := (i_1, i_2, i_3)$.

Now we prove, for the integral operators introduced, some theorems which are the exact structural analogues of the corresponding facts of one-dimensional complex analysis and which express very profound properties of $\alpha$-holomorphic function theory, as well as very important relations between this theory and operator theory.

**4.11 Theorem** (Quaternionic Borel–Pompeiu formula for a complex parameter $\alpha$) *Let $\Omega$ be a domain in $\mathbb{R}^3$ with the Liapunov boundary $\Gamma := \partial\Omega$. Let $\alpha_0 \in \mathbb{C}$ and $f \in C^1(\Omega; \mathbb{H}(\mathbb{C})) \cap C(\bar{\Omega}; \mathbb{H}(\mathbb{C}))$. Then*

$$ {}^{\psi}K_{\alpha_0}[f](x) + {}^{\psi}T_{\alpha_0} \cdot {}^{\psi}D_{\alpha_0}[f](x) = f(x), \quad x \in \Omega. \tag{4.13}$$

PROOF. We have

$$ {}^{\psi}T_{\alpha_0} \cdot {}^{\psi}D_{\alpha_0}[f](x) = \int_{\Omega} \mathcal{K}_{\psi,\alpha_0}(x-y){}^{\psi}D_{\alpha_0,y}[f](y)dy = $$

$$= \lim_{\epsilon\to 0}\int_{\Omega^\epsilon} \mathcal{K}_{\psi,\alpha_0}(x-y)\,{}^\psi D_{\alpha_0,y}[f](y)dy, \tag{4.14}$$

where the index $y$ in ${}^\psi D_{\alpha_0,y}$ means differentiation with respect to $y$;

$$\Omega^\epsilon := \Omega\backslash\{y \,|\, |x-y| \le \epsilon\} = \Omega\backslash B^\epsilon(x).$$

Let us transform the integral

$$\int_{\Omega^\epsilon} \mathcal{K}_{\psi,\alpha_0}(x-y)\,{}^\psi D_{\alpha_0,y}[f](y)dy =$$

$$= \int_{\Omega^\epsilon} \mathcal{K}_{\psi,\alpha_0}(x-y)\,{}^\psi D_y[f](y)dy + \alpha_0\int_{\Omega^\epsilon} \mathcal{K}_{\psi,\alpha_0}(x-y)f(y)dy =$$

$$= \int_{\Omega^\epsilon} \mathcal{K}_{\psi,\alpha_0}(x-y)\,{}^\psi D_y[f](y)dy + \alpha_0\int_{\Omega^\epsilon} \mathcal{K}_{\psi,\alpha_0}(x-y)f(y)dy -$$

$$-\int_{\Omega^\epsilon} D_y^\psi[\mathcal{K}_{\psi,\alpha_0}(x-y)]f(y)dy + \int_{\Omega^\epsilon} D_y^\psi[\mathcal{K}_{\psi,\alpha_0}(x-y)]f(y)dy. \tag{4.15}$$

By definition of $\mathcal{K}_{\psi,\alpha_0}$,

$$-D_y^\psi[\mathcal{K}_{\psi,\alpha_0}(x-y)] = -{}^\psi D_y[\mathcal{K}_{\psi,\alpha_0}(x-y)] = {}^\psi D_x[\mathcal{K}_{\psi,\alpha_0}(x-y)].$$

Further, applying (4.1), we see that

$$\int_{\Omega^\epsilon} \mathcal{K}_{\psi,\alpha_0}(x-y)\,{}^\psi D_{\alpha_0,y}[f](y)dy =$$

$$= \int_{\Omega^\epsilon}(\mathcal{K}_{\psi,\alpha_0}(x-y)\,{}^\psi D_y[f(y)] + D_y^\psi[\mathcal{K}_{\psi,\alpha_0}(x-y)])f(y))dy+$$

$$+\int_{\Omega^\epsilon}(\alpha_0\mathcal{K}_{\psi,\alpha_0}(x-y) + {}^\psi D_x[\mathcal{K}_{\psi,\alpha_0}(x-y)])f(y)dy =$$

$$= \int_{\Gamma^\epsilon} \mathcal{K}_{\psi,\alpha_0}(x-\tau)\vec{n}_\psi(\tau)f(\tau)d\Gamma^\epsilon_\tau + \int_{\Omega^\epsilon} {}^\psi D_{\alpha_0}[\mathcal{K}_{\psi,\alpha_0}(x-y)]f(y)dy, \tag{4.16}$$

where $\Gamma^\epsilon := \partial\Omega^\epsilon$. Passing to the limit in (4.16) we obtain (4.13). □

This theorem immediately implies the following

**4.12 Theorem** (Quaternionic Cauchy integral formula for a complex parameter $\alpha_0$) *Let $\Omega$ be a domain in $\mathbb{R}^3$ with the Liapunov boundary $\Gamma := \partial\Omega$. Let $f \in {}^\psi\mathfrak{M}_{\alpha_0}(\Omega) \cap C(\bar{\Omega})$. Then*

$$f(x) = {}^\psi K_{\alpha_0}[f](x), \quad x \in \Omega. \tag{4.17}$$

Now we prove an inverse (in a certain sense) theorem to the quaternionic Cauchy theorem 4.9, that is, an analogue of the complex Morera theorem.

**4.13 Theorem** (Quaternionic Morera theorem for a complex parameter $\alpha_0$) *Let $\alpha_0 \in \mathbb{C}$, $f \in C^1(\Omega; \mathbb{H}(\mathbb{C}))$, ${}^\psi D_{\alpha_0}[f] \in L_p(\Omega; \mathbb{H}(\mathbb{C}))$ for some $p > 1$. If for any Liapunov manifold-without-boundary $\hat{\Gamma}$ $(\hat{\Gamma} = \partial\hat{\Omega}, \hat{\Omega} \subset \Omega, \hat{\Gamma} \subset \Omega)$ the following equality holds,*

$$\int_{\hat{\Gamma}} \vec{n}_\psi f d\hat{\Gamma} = -\int_{\hat{\Omega}} f\alpha_0 dx, \tag{4.18}$$

*then $f$ is $(\psi, \alpha_0)$-holomorphic in $\Omega$.*

PROOF. Let $\{\Omega_k\}_{k\in\mathbb{N}}$ be a regular sequence of domains converging to the fixed point $x_0 \in \Omega$. We denote $\partial\Omega_k$ by $\Gamma_k$. By Lebesgue's theorem (see, e.g., [121]) for any $\mathbb{H}(\mathbb{C})$-valued function $g \in L_p(\Omega)$, $p > 1$, there exists

$$\lim_{k\to\infty} \frac{1}{|\Omega_k|} \int_{\Omega_k} g dx =: \tilde{g}(x_0),$$

and $g = \tilde{g}$ in $L_p(\Omega)$. Let us choose $g := {}^\psi D[f]$. By the formula (4.1) (when $\alpha = 0$ in it) we have

$$\int_{\Omega_k} {}^\psi D[f]dx = \int_{\Gamma_k} \vec{n}_\psi f d\Gamma_k.$$

From the hypothesis of the theorem, it follows that

$$\frac{1}{|\Omega_k|}\int_{\Omega_k} {}^{\psi}\!\mathcal{D}_{\alpha_0}[f]dx = 0 \quad \text{for} \quad k \in \{0\} \cup \mathbb{N}.$$

Therefore ${}^{\psi}\!\mathcal{D}_{\alpha_0}[f](x) = 0$ almost everywhere in $\Omega$. □

**4.14 Theorem** (Right inverse for the quaternionic Cauchy–Riemann operator for the case of a complex parameter $\alpha_0$) *Let* $\alpha_0 \in \mathbb{C}$, *and* $f \in C^1(\Omega; \mathbb{H}(\mathbb{C})) \cap C(\bar{\Omega}; \mathbb{H}(\mathbb{C}))$. *Then the following equality holds*

$$ {}^{\psi}\!\mathcal{D}_{\alpha_0} \cdot {}^{\psi}T_{\alpha_0}[f](x) = f(x), \quad \forall x \in \Omega. \tag{4.19}$$

PROOF. Firstly let us remark that from the quaternionic Stokes formula (4.1) we have the equality

$$\int_{\Gamma} \theta_{\alpha_0}(x-\tau)\vec{n}_{\psi}(\tau)f(\tau)d\Gamma_\tau = \int_{\Omega}\{\mathcal{K}_{\psi,\alpha_0}(x-y)f(y) - \theta_{\alpha_0}(x-y)^{\bar{\psi}}D_{\alpha_0}[f(y)]\}dx,$$

which implies

$${}^{\psi}T_{\alpha_0}[f](x) = \int_{\Omega} \theta_{\alpha_0}(x-y)^{\bar{\psi}}D_{\alpha_0}[f](y)dx + \int_{\Gamma}\theta_{\alpha_0}(x-\tau)\vec{n}_{\psi}(\tau)f(\tau)d\Gamma_\tau .$$

Applying the operator ${}^{\psi}\!\mathcal{D}_{\alpha_0}$ and using the Borel–Pompeiu formula (4.13) we obtain:

$${}^{\psi}\!\mathcal{D}_{\alpha_0} \cdot {}^{\psi}T_{\alpha_0}[f](x) =$$

$$= -\int_{\Omega} {}^{\psi}\!\mathcal{D}_{\alpha_0}[\theta_{\alpha_0}(x-y)]^{\psi}\mathcal{D}_{-\alpha}[f](y)dy + \int_{\Gamma} {}^{\psi}\!\mathcal{D}_{\alpha_0}[\theta_{\alpha_0}(x-\tau)]\vec{n}_{\psi}(\tau)f(\tau)d\Gamma_\tau =$$

$$= \int_{\Omega} \mathcal{K}_{\psi,-\alpha_0}(x-y)^{\psi}\mathcal{D}_{-\alpha_0}[f](y)dy - \int_{\Gamma}\mathcal{K}_{\psi,-\alpha_0}(x-\tau)\vec{n}_{\psi}(\tau)f(\tau)d\Gamma_\tau =$$

$$= ({}^{\psi}T_{-\alpha_0}\,{}^{\psi}\mathcal{D}_{-\alpha_0} + {}^{\psi}K_{-\alpha_0})[f](x) = f(x).$$

$\square$

**4.15** The quaternionic Borel–Pompeiu formula, the Cauchy integral formula, the Morera theorem and the theorem on the right inverse for the operator ${}^{\psi}\mathcal{D}_{\alpha}$ were proved under the assumption $\alpha = \alpha_0 \in \mathbb{C}$. Now we show that the integral operators associated with ${}^{\psi}\mathcal{D}_{\alpha_0}$ can be introduced for an arbitrary $\alpha \in \mathbb{H}(\mathbb{C})$ in such a way that all the above-mentioned facts remain true. The form of the fundamental solution $\mathcal{K}_{\psi,\alpha}$ (Theorem 3.16) suggests how to accomplish this.

We extend the definition of the integral operators introduced above ${}^{\psi}T_{\alpha}$, ${}^{\psi}K_{\alpha}$, ${}^{\psi}S_{\alpha}$ for $\alpha \in \mathbb{H}(\mathbb{C})$ as follows:

$$
{}^{\psi}T_{\alpha} := \begin{cases} P^{+} \cdot {}^{\psi}T_{\xi_{+}} + P^{-} \cdot {}^{\psi}T_{\xi_{-}}, & \alpha \notin \mathfrak{S}, \vec{\alpha}^{2} \neq 0, \\ \\ {}^{\psi}T_{\alpha_0} + M^{\vec{\alpha}} \frac{\partial}{\partial \alpha_0}[{}^{\psi}T_{\alpha_0}], & \alpha \notin \mathfrak{S}, \vec{\alpha}^{2} = 0, \\ \\ P^{+} \cdot {}^{\psi}T_{2\alpha_0} + P^{-} \cdot {}^{\psi}T_{0}, & \alpha \in \mathfrak{S}, \alpha_0 \neq 0, \\ \\ {}^{\psi}T_{0} + M^{\alpha} \cdot W_{0}, & \alpha \in \mathfrak{S}, \alpha_0 = 0, \end{cases} \tag{4.20}
$$

where

$$
W_{\mu}[f](x) := \int_{\Omega} \theta_{\mu}(x-y) f(y) dy, \mu \in \mathbb{C}, \quad x \in \mathbb{R}^{3},
$$

$$
{}^{\psi}K_{\alpha} := \begin{cases} P^{+} \cdot {}^{\psi}K_{\xi_{+}} + P^{-}\, {}^{\psi}K_{\xi_{-}}, & \alpha \notin \mathfrak{S}, \vec{\alpha}^{2} \neq 0, \\ \\ {}^{\psi}K_{\alpha_0} + M^{\vec{\alpha}} \frac{\partial}{\partial \alpha_0}[{}^{\psi}K_{\alpha_0}], & \alpha \notin \mathfrak{S}, \vec{\alpha}^{2} = 0, \\ \\ P^{+} \cdot {}^{\psi}K_{2\alpha_0} + P^{-} \cdot {}^{\psi}K_{0}, & \alpha \in \mathfrak{S}, \alpha_0 \neq 0, \\ \\ {}^{\psi}K_{0} - M^{\alpha} \cdot {}^{\psi}V_{0}, & \alpha \in \mathfrak{S}, \alpha_0 = 0, \end{cases} \tag{4.21}
$$

where

$$ {}^{\psi}V_\mu[f](x) := \int_\Gamma \theta_\mu(x-\tau)\cdot\vec{n}_\psi(\tau)f(\tau)d\Gamma_\tau, \mu\in\mathbb{C}, x\in\mathbf{R}^3\backslash\Gamma, $$

$$ {}^{\psi}S_\alpha := \begin{cases} P^+\cdot{}^{\psi}S_{\xi_+} + P^-\cdot{}^{\psi}S_{\xi_-}, & \alpha\notin\mathfrak{S}, \vec{\alpha}^2\neq 0, \\ {}^{\psi}S_{\alpha_0} + M^{\vec{\alpha}}\frac{\partial}{\partial\alpha_0}[{}^{\psi}S_{\alpha_0}], & \alpha\notin\mathfrak{S}, \vec{\alpha}^2 = 0, \\ P^+\cdot{}^{\psi}S_{2\alpha_0} + P^-\cdot{}^{\psi}S_0, & \alpha\in\mathfrak{S}, \alpha_0\neq 0, \\ S_0 - M^\alpha\cdot{}^{\psi}\hat{V}_0, & \alpha\in\mathfrak{S}, \alpha_0 = 0, \end{cases} \tag{4.22} $$

and

$$ {}^{\psi}\hat{V}_\mu[f](x) := 2\int_\Gamma \theta_\mu(x-\tau)\vec{n}_\psi(\tau)f(\tau)d\Gamma_\tau, \mu\in\mathbb{C}, x\in\Gamma. $$

It is easy to verify that in the special case $\alpha=\alpha_0\in\mathbb{C}$ the operators ${}^{\psi}T_\alpha, {}^{\psi}K_\alpha, {}^{\psi}S_\alpha$ coincide with those introduced in Subsection 4.10. We define as before the operators ${}^{\psi}P_\alpha$ and ${}^{\psi}Q_\alpha$ by (4.12), where ${}^{\psi}S_\alpha$ is already the operator (4.22).

Let us remark that we can propose another subdivision into cases. Namely, we can consider the case $\alpha\in\mathbb{H}(\mathbb{C})$, $\alpha_0=0$. Then it can be easily verified that for $\alpha\notin\mathfrak{S}$ and for $\alpha\in\mathfrak{S}$ the operator of singular integration, for example, has the same structure:

$$ {}^{\psi}S_\alpha[f](x) = \begin{cases} (P^+\,{}^{\psi}S_{\xi_+} + P^-\cdot{}^{\psi}S_{\xi_-})[f](x), & \vec{\alpha}\notin\mathfrak{S} \\ (S_0 - M^{\vec{\alpha}}\,{}^{\psi}\hat{V}_0)[f](x), & \vec{\alpha}\in\mathfrak{S} \end{cases} = $$

$$ = -2\int_\Gamma\{\theta_\gamma(x-\tau)(\frac{(x-\tau)_\psi}{|x-\tau|^2} + i\gamma\frac{(x-\tau)_\psi}{|x-\tau|})\vec{n}_\psi(\tau)f(\tau)+ $$

$$+\theta_\gamma(x-\tau)\vec{n}_\psi(\tau)f(\tau)\vec{\alpha}\}d\Gamma_\tau, \quad x \in \Gamma. \tag{4.23}$$

Here, as before, $\gamma \in \mathbb{C}$, $\gamma^2 = \vec{\alpha}^2$; $(x-\tau)_\psi := \sum_{k=1}^3 \psi^k(x_k - \tau_k)$.

Note, that we did not diminish the number of different cases under consideration. The representation (4.23) allows only another way of subdividing them.

**4.16** Remark We can rewrite the operators (4.20)–(4.22) in a form similar to (4.9)–(4.11) but then the Cauchy kernel $\mathcal{K}_{\psi,\alpha}$, a function, must give up its place in favour of an operator-valued kernel. For example, the operator ${}^\psi\mathcal{K}_\alpha$ can be defined as follows:

$${}^\psi\mathcal{K}_\alpha[f](x) = -\int_\Gamma {}^\psi\check{\mathcal{K}}_\alpha^x[\vec{n}_\psi \cdot f](\tau)d\Gamma_\tau, \quad x \in \mathbb{R}^3\backslash\Gamma, \tag{4.24}$$

where the operator ${}^\psi\check{\mathcal{K}}_\alpha$ is given by:

$$\check{\mathcal{K}}_\alpha^x[g](\tau) := \begin{cases} ({}^\psi\mathcal{K}_{\xi_+}(x-\tau)P^+ + {}^\psi\mathcal{K}_{\xi_-}(x-\tau)P^-)[g](\tau), & \alpha \notin \mathfrak{S}, \vec{\alpha}^2 \neq 0, \\ ({}^\psi\mathcal{K}_{\alpha_0}(x-\tau)I + \frac{\partial}{\partial\alpha_0}[{}^\psi\mathcal{K}_{\alpha_0}(x-\tau)]M^{\vec{\alpha}})[g](\tau), & \alpha \notin \mathfrak{S}, \vec{\alpha}^2 = 0, \\ ({}^\psi\mathcal{K}_{2\alpha_0}(x-\tau)P^+ + {}^\psi\mathcal{K}_0(x-\tau)P^-)[g](\tau), & \alpha \in \mathfrak{S}, \alpha_0 \neq 0, \\ ({}^\psi\mathcal{K}_0(x-\tau)I + \theta_0(x-\tau)M^\alpha)[g](\tau), & \alpha \in \mathfrak{S}, \alpha_0 = 0. \end{cases} \tag{4.25}$$

By analogy, we can rewrite:

$${}^\psi T_\alpha[f](x) = \int_\Omega {}^\psi\check{\mathcal{K}}_\alpha^x[f](y)dy, \quad x \in \mathbb{R}^3, \tag{4.26}$$

$${}^\psi S_\alpha[f](x) = -2\int_\Gamma {}^\psi\check{\mathcal{K}}_\alpha^x[\vec{n}_\psi \cdot f](\tau)d\Gamma_\tau, \quad x \in \Gamma. \tag{4.27}$$

Thus, defining the operators ${}^{\psi}T_\alpha$, ${}^{\psi}K_\alpha$, ${}^{\psi}S_\alpha$ for $\alpha \in \mathbb{H}(\mathbb{C})$ we come across a new phenomenon, where the Cauchy kernel becomes an operator containing the multiplications from both sides. It is as if the Cauchy kernel is spread to the whole integral.

We have not yet proved that the operators ${}^{\psi}T_\alpha$, ${}^{\psi}K_\alpha$, ${}^{\psi}S_\alpha$, $\alpha \in \mathbb{H}(\mathbb{C})$, are really analogues of the $T$-operator, the Cauchy integral operator and the operator of singular integration; we must prove the corresponding properties of these operators. We start with the following.

**4.17 Theorem** (Main integral theorems for $\alpha$-holomorphic functions in the case of an arbitrary quaternionic parameter $\alpha$) *Let $\alpha$ be an arbitrary complex quaternion, $\Omega$ be a domain in $\mathbb{R}^3$ with the Liapunov boundary $\Gamma := \partial\Omega$, and let ${}^{\psi}T_\alpha$, ${}^{\psi}K_\alpha$ and ${}^{\psi}S_\alpha$ be as in formulas (4.20)–(4.22). Then the following assertions are true:*

1) *(quaternionic Borel–Pompeiu formula for a complex-quaternionic parameter $\alpha$).*
*If $f \in C^1(\Omega; \mathbb{H}(\mathbb{C})) \cap C(\bar{\Omega}; \mathbb{H}(\mathbb{C}))$ then $\forall x \in \Omega$*

$$ {}^{\psi}K_\alpha[f](x) + {}^{\psi}T_\alpha \cdot {}^{\psi}D_\alpha[f](x) = f(x); $$

2) *(quaternionic Cauchy integral formula for a complex-quaternionic parameter $\alpha$).*
*If $f \in {}^{\psi}\mathfrak{M}_\alpha(\Omega, \mathbb{H}(\mathbb{C})) \cap C(\bar{\Omega}; \mathbb{H}(\mathbb{C}))$ then $\forall x \in \Omega$*

$$ f(x) = {}^{\psi}K_\alpha[f](x). $$

3) *(quaternionic Morera theorem for a complex-quaternionic parameter $\alpha$).*

*If $f \in C^1(\Omega; \mathbb{H}(\mathbb{C}))$, ${}^{\psi}D_\alpha[f] \in L_p(\Omega; \mathbb{H}(\mathbb{C}))$, $p > 1$, and if for any Liapunov manifold-without-boundary $\gamma$ $(\gamma = \partial\Xi, \Xi \subset \Omega, \gamma \in \Omega)$ the following equality holds:*

$$ \int_\gamma n_\psi f \, d\gamma = -\int_\Xi f \alpha \, dx, $$

*then $f$ is $(\psi, \alpha)$– holomorphic in $\Omega$.*

4) *(right inverse for the quaternionic Cauchy–Riemann operator for a complex–quaternionic parameter* $\alpha$*). If* $f \in C^1(\Omega, \mathbb{H}(\mathbb{C})) \cap C(\bar{\Omega}; \mathbb{H}(\mathbb{C}))$ *then* $\forall x \in \Omega$

$$^{\psi}\!\mathcal{D}_\alpha \cdot {}^{\psi}T_\alpha[f](x) = f(x).$$

**Proof.** It is clear that the most essential fact, in the course of the proof, is that the definitions of operators (4.20)–(4.22) contain operators ${}^{\psi}T_\mu, {}^{\psi}K_\mu, {}^{\psi}S_\mu$, with $\mu \in \mathbb{C}$ for which corresponding results have already been obtained.

1) We begin with the first assertion. We are forced to consider four cases separately. Let us first examine the case $\alpha \notin \mathfrak{S}, \vec{\alpha}^2 \neq 0$. We have

$$^{\psi}\!\mathcal{K}_\alpha[f] = P^+[f - {}^{\psi}T_{\xi_+}\ {}^{\psi}\!\mathcal{D}_{\xi+}[f]] + P^-[f - {}^{\psi}T_{\xi_-} \cdot {}^{\psi}D_{\xi_-}[f]] =$$

$$= f - {}^{\psi}T_{\xi_+} P^+\ {}^{\psi}\!\mathcal{D}_{\xi_+}[f] - {}^{\psi}T_{\xi_-} P^- \cdot {}^{\psi}D_{\xi_-}[f].$$

From the permutability of the operators $P^\pm$ with $D_{\xi\pm}$ and from (3.6) we see

$$P^+\ {}^{\psi}\!\mathcal{D}_{\xi+}[f] = P^+\ {}^{\psi}\!\mathcal{D}_\alpha[f]; \quad P^-\ {}^{\psi}\!\mathcal{D}_{\xi-}[f] = P^-\ {}^{\psi}\!\mathcal{D}_\alpha[f]. \tag{4.28}$$

Therefore we obtain the Borel–Pompeiu formula:

$$^{\psi}\!\mathcal{K}_\alpha[f] = f - (P^{+\psi}T_{\xi+} + P^-\ {}^{\psi}T_{\xi-})\ {}^{\psi}\!\mathcal{D}_\alpha[f] = f - {}^{\psi}T_\alpha \cdot {}^{\psi}D_\alpha[f].$$

For the cases 2), 3) the proof is similar.

For the case $\alpha \in \mathfrak{S}, \alpha_0 = 0$ it is necessary to use in addition the equality

$$^{\psi}V_0[f](x) = \int_\Omega {}^{\psi}\!\mathcal{D}[\theta_0(x-y)f(y)]dy,$$

which follows directly from the formula (4.3) (the differentiation here is with respect to $y$). Then, as ${}^{\psi}\!\mathcal{D}[\theta_0(x-y)] = {}^{\psi}\!\mathcal{K}_0(x-y)$ we have (using the quaternionic Leibnitz differentiation rule (3.12)):

$$ {}^{\psi}\mathcal{D}[\theta_0(x-y)f(y)] = {}^{\psi}\mathcal{K}_0(x-y)f(y) + \theta_0(x-y){}^{\psi}\mathcal{D}[f(y)]. $$

And therefore

$$ {}^{\psi}V_0[f](x) = {}^{\psi}T_0[f](x) + W_0 \cdot {}^{\psi}D[f](x). \tag{4.29} $$

Consequently,

$$ {}^{\psi}K_\alpha[f] = f - {}^{\psi}T_0 \cdot {}^{\psi}D[f] - W_0{}^{\psi}D_\alpha[f]\alpha - {}^{\psi}T_0[f]\alpha = f - {}^{\psi}T_\alpha \cdot {}^{\psi}D_\alpha[f], $$

and the quaternionic Borel–Pompeiu formula for $\alpha \in \mathbb{H}(\mathbb{C})$ is proved.
2)–3). Both the quaternionic Cauchy integral formula and the Morera theorem do not require new reasoning.
4) To prove the assertion about a right inverse for the Cauchy–Riemann operator, let us consider, first, the case $\alpha \in \mathfrak{S}, \vec{\alpha}^2 \neq 0$:

$$ \begin{aligned} {}^{\psi}\mathcal{D}_\alpha\, {}^{\psi}T_\alpha[f] &= ({}^{\psi}\mathcal{D}_{\xi_+}P^+ + {}^{\psi}D_{\xi_-}P^-)(P^+ \cdot {}^{\psi}T_{\xi_+} + + P^- \cdot {}^{\psi}T_{\xi_-})[f] = \\ &= (P^+ \cdot {}^{\psi}D_{\xi_+} \cdot {}^{\psi}T_{\xi_+} + P^-\, {}^{\psi}\mathcal{D}_{\xi_-}\, {}^{\psi}T_{\xi_-})[f] = \\ &= (P^+ + P^-)[f] = f. \end{aligned} $$

We have made use of (3.6) and of the formula (4.19) for complex parameters $\xi_\pm$.

The proof for the case $\alpha \in \mathfrak{S}$, $\alpha_0 \neq 0$ is similar.

Let $\alpha \in \mathfrak{S}$, $\alpha_0 = 0$, then

$$ {}^{\psi}\mathcal{D}_\alpha{}^{\psi}T_\alpha[f] = ({}^{\psi}\mathcal{D}_\alpha{}^{\psi}T_0 + {}^{\psi}D_\alpha M^\alpha W_0)[f] = f + M^\alpha\, {}^{\psi}T_0[f] + M^\alpha\, {}^{\psi}\mathcal{D}W_0[f]. $$

But ${}^{\psi}\mathcal{D}W_0 = -{}^{\psi}T_0$, hence for $\alpha \in \mathfrak{S}$, $\alpha_0 = 0$ the formula (4.19) is valid.
We now consider the last case, $\alpha \notin \mathfrak{S}$, $\vec{\alpha}^2 = 0$.

$$ {}^{\psi}\mathcal{D}_\alpha{}^{\psi}T_\alpha[f] = {}^{\psi}D_\alpha{}^{\psi}T_{\alpha_0}[f] + {}^{\psi}D_\alpha \frac{\partial}{\partial \alpha_0}[{}^{\psi}T_{\alpha_0}[f]]\vec{\alpha} = $$

$$= f + {}^{\psi}T_{\alpha_0}[f]\vec{\alpha} + {}^{\psi}D_{\alpha_0}\frac{\partial}{\partial\alpha_0}[{}^{\psi}T_{\alpha_0}[f]]\vec{\alpha} =$$

$$= f + ({}^{\psi}T_{\alpha_0}[f])\vec{\alpha} + \frac{\partial}{\partial\alpha_0}({}^{\psi}\mathcal{D}_{\alpha_0}{}^{\psi}T_{\alpha_0}[f])\vec{\alpha} - {}^{\psi}T_{\alpha_0}[f]\vec{\alpha} =$$

$$= f + \frac{\partial}{\partial\alpha_0}(f)\vec{\alpha} = f.$$

□

**4.18 Theorem** (Factorization of the operator ${}^{\psi}\mathcal{K}_{\alpha}$) *The operator ${}^{\psi}\mathcal{K}_{\alpha}$ admits the following representation:*

$$ {}^{\psi}\mathcal{K}_{\alpha} = \begin{cases} (4\alpha_0\gamma)^{-1}\,{}^{\psi}\mathcal{D}_{\vec{\alpha}}\,{}^{\psi}\mathcal{D}_{-\alpha}\,{}^{\psi}\mathcal{D}_{-\vec{\alpha}}({}^{\psi}V_{\xi_-} - {}^{\psi}V_{\xi_+}), & \alpha \notin \mathfrak{S},\quad \vec{\alpha}^2 \neq 0, \alpha_0 \neq 0, \\ {}^{\psi}\mathcal{D}_{-\alpha}\cdot{}^{\psi}V_{\gamma}, & \alpha \notin \mathfrak{S},\quad \alpha_0 = 0, \\ (I + M^{\vec{\alpha}}\frac{\partial}{\partial\alpha_0})\,{}^{\psi}\mathcal{D}_{-\alpha_0}\cdot{}^{\psi}V_{\alpha_0}, & \alpha \notin \mathfrak{S},\quad \vec{\alpha}^2 = 0, \\ P^+\cdot{}^{\psi}D_{-2\alpha_0}\cdot{}^{\psi}V_{2\alpha_0} + P^-\cdot{}^{\psi}D\cdot{}^{\psi}V_0, & \alpha \in \mathfrak{S},\quad \alpha_0 \neq 0, \\ {}^{\psi}\mathcal{D}_{-\alpha}\cdot{}^{\psi}V_0, & \alpha \in \mathfrak{S},\quad \alpha_0 = 0. \end{cases} \tag{4.30}$$

PROOF. For all cases but the first one ($\alpha \notin \mathfrak{S}$, $\vec{\alpha}^2 \neq 0$, $\alpha_0 \neq 0$), the representation (4.30) follows immediately from the definition of the function ${}^{\psi}\mathcal{K}_{\alpha}$, $\alpha \in \mathbb{C}$ (see Theorem 3.16).

Let us examine the first case. For an $\mathbb{H}(\mathbb{C})$-valued function $f$ continuous on $\Gamma$, we have

$${}^{\psi}\mathcal{D}_{\vec{\alpha}}\,{}^{\psi}\mathcal{D}_{-\alpha}\,{}^{\psi}\mathcal{D}_{-\vec{\alpha}}({}^{\psi}V_{\xi_-} - {}^{\psi}V_{\xi_+})[f] = ({}^{\psi}\mathcal{D}_{\vec{\alpha}}(-\Delta({}^{\psi}V_{\xi_-} - {}^{\psi}V_{\xi_+}) - M^{\vec{\alpha}}\,{}^{\psi}\mathcal{D}({}^{\psi}V_{\xi_-} - {}^{\psi}V_{\xi_+}) -$$

$$- M^{\alpha}\,{}^{\psi}\mathcal{D}({}^{\psi}V_{\xi_-} - {}^{\psi}V_{\xi_+}) + |\alpha|^2({}^{\psi}V_{\xi_-} - {}^{\psi}V_{\xi_+})))[f].$$

The simple layer potential ${}^{\psi}V_{\mu}[f]$ satisfies the Helmholtz equation $(\Delta + \mu^2 I)$ $[{}^{\psi}V_{\mu}[f]] = 0$. Thus we obtain

$$ {}^{\psi}\mathcal{D}_{\bar{\alpha}}\ {}^{\psi}\mathcal{D}_{-\alpha}\ {}^{\psi}\mathcal{D}_{-\bar{\alpha}}({}^{\psi}V_{\xi_-} - {}^{\psi}V_{\xi_+})[f] = $$

$$ = {}^{\psi}\mathcal{D}_{\bar{\alpha}}(\xi_-^2\,{}^{\psi}V_{\xi_-} - \xi_+^2\,{}^{\psi}V_{\xi_+} - 2\alpha_0{}^{\psi}D({}^{\psi}V_{\xi_-} - {}^{\psi}V_{\xi_+}) + |\alpha|^2({}^{\psi}V_{\xi_-} - {}^{\psi}V_{\xi_+}))[f]. $$

Note that $|\alpha|^2 = \alpha_0^2 - \vec{\alpha}^2 = \alpha_0^2 - \gamma^2 = \xi_+\xi_-$. Hence

$$ {}^{\psi}\mathcal{D}_{\bar{\alpha}}\ {}^{\psi}\mathcal{D}_{-\alpha}\ {}^{\psi}\mathcal{D}_{-\bar{\alpha}}({}^{\psi}V_{\xi_-} - {}^{\psi}V_{\xi_+})[f] = $$

$$ = {}^{\psi}D_{\bar{\alpha}}(\xi_-(\xi_+ + \xi_-){}^{\psi}V_{\xi_-} - \xi_+(\xi_+ + \xi_-){}^{\psi}V_{\xi_+} - 2\alpha_0\,{}^{\psi}\mathcal{D}({}^{\psi}V_{\xi_-} - {}^{\psi}V_{\xi_+}))[f] = $$

$$ = ((\xi_+ + \xi_-)(\xi_-\ {}^{\psi}\mathcal{D}\ {}^{\psi}V_{\xi_-} - \xi_+\ {}^{\psi}\mathcal{D}\ {}^{\psi}V_{\xi_+}) + $$

$$ +2\alpha_0\Delta({}^{\psi}V_{\xi_-} - {}^{\psi}V_{\xi_+}) + (\xi + \xi_-)(\xi_-\ {}^{\psi}V_{\xi_-} - \xi_+\ {}^{\psi}V_{\xi_+})M^{\bar{\alpha}} - $$

$$ -2\alpha_0\ {}^{\psi}\mathcal{D}({}^{\psi}V_{\xi_-} - {}^{\psi}V_{\xi_+})M^{\bar{\alpha}})[f]. $$

We calculate:

$$ (2\alpha_0\Delta{}^{\psi}V_{\xi_-} + (\xi_+ + \xi_-)\xi_-\ {}^{\psi}V_{\xi_-}M^{\bar{\alpha}})[f] = \xi_-\ {}^{\psi}V_{\xi_-}(-2\alpha_0\xi_- I + (\xi_+ + \xi_-)M^{\alpha})[f] = $$

$$ = \xi_-\ {}^{\psi}V_{\xi_-}(\xi_+ M^{\bar{\alpha}} - \xi_- M^{\alpha})[f] = 2\alpha_0\xi_-\ {}^{\psi}V_{\xi_-}M^{(\gamma - \vec{\alpha})}[f]. $$

Further,

$$ -2\alpha_0\vec{\alpha} + \xi_+\xi_- + \xi_-^2 = -2\alpha_0^2 + 2\alpha_0\vec{\alpha} + \alpha_0^2 - \gamma^2 + \alpha_0^2 + \gamma^2 - 2\alpha_0\gamma = -2\alpha_0(\gamma - \vec{\alpha}). $$

Similarly,

$$2\alpha_0\vec{\alpha} - \xi_+^2 - \xi_-\xi_+ = -2\alpha_0(\gamma + \vec{\alpha}),$$

and

$$(-2\alpha_0\Delta {}^{\psi}V_{\xi+} - \xi_+(\xi_+ + \xi_-){}^{\psi}V_{\xi+}M^{\vec{\alpha}})[f] = -2\alpha_0\xi_+ \, {}^{\psi}V_{\xi+}M^{(\gamma+\vec{\alpha})}[f].$$

Thus we finally obtain

$$(4\alpha_0\gamma)^{-1} \, {}^{\psi}\mathcal{D}_{\bar{\alpha}} \, {}^{\psi}\mathcal{D}_{-\alpha} \, {}^{\psi}\mathcal{D}_{-\bar{\alpha}}({}^{\psi}V_{\xi-} - \, {}^{\psi}V_{\xi+}) =$$

$$= (4\alpha_0\gamma)^{-1}(2\alpha_0\xi_- \, {}^{\psi}V_{\xi-}M^{(\gamma-\vec{\alpha})} - 2\alpha_0 \, {}^{\psi}\mathcal{D}{}^{\psi}V_{\xi-}M^{(\gamma-\vec{\alpha})}$$

$$+2\alpha_0\xi_+ \, {}^{\psi}V_{\xi+}M^{(\gamma+\vec{\alpha})} - 2\alpha_0 \, ({}^{\psi}\mathcal{D}{}^{\psi}V_{\xi+}M^{(\gamma+\vec{\alpha})}) =$$

$$= P^+(-{}^{\psi}\mathcal{D}{}^{\psi}V_{\xi+} + \xi_+^{\psi}V_{\xi+}) + P^-(-{}^{\psi}\mathcal{D}{}^{\psi}V_{\xi-} + \xi_- \, {}^{\psi}V_{\xi-}) =$$

$$= P^+ \cdot {}^{\psi}K_{\xi+} + P^- \cdot {}^{\psi}K_{\xi-} = {}^{\psi}K_{\alpha}.$$

□

**4.19** Now we give some first applications of Theorem 4.17 to $\lambda$-metaharmonic functions for $\lambda \in \mathbb{H}(\mathbb{C})$. Based on the decomposition Theorems 2.13 and 2.18, and on assertion 1) of Theorem 4.17, we shall obtain an integral representation of an arbitrary function $f \in \ker\Delta_\lambda$ from its boundary values. We start with the harmonic case, $\lambda = 0$.

**4.20 Theorem** *Let* $f \in C^2(\Omega) \cap C^1(\bar{\Omega})$. *Then* $f \in \ker\Delta$ *iff*

$$f = {}^{\psi}K_0[f] + {}^{\psi}V_0 \, {}^{\psi}\mathcal{D}[f]. \tag{4.31}$$

PROOF. From Theorem 2.18 we have:

$$f \in \ker\Delta \iff f = {}^{\psi}\Pi_\alpha^+[f] + {}^{\psi}\Pi_\alpha^-[f]$$

where $\alpha \in \mathfrak{S}$, $\alpha_0 = 0$, $\alpha_1 \neq 0$, and the projection operators ${}^{\psi}\Pi_\alpha^\pm$ are defined by (2.41), ${}^{\psi}\Pi_\alpha^+[f] \in {}^{\psi}\mathfrak{M}_{-\alpha}$, ${}^{\psi}\Pi_\alpha^-[f] \in M^{i_1}({}^{\psi}\mathfrak{M}_\alpha)$. Taking this into account we can now use the quaternionic Cauchy formula (Theorem 4.17) for each of the classes ${}^{\psi}\mathfrak{M}_{\pm\alpha}$, which gives

$${}^{\psi}\Pi_\alpha^+[f] = ({}^{\psi}\mathcal{K}_0 + M^\alpha \cdot {}^{\psi}V_0)[{}^{\psi}\Pi_\alpha^+[f]];$$

$${}^{\psi}\Pi_\alpha^-[f] \cdot i_1 = ({}^{\psi}\mathcal{K}_0 - M^\alpha\, {}^{\psi}V_0)[{}^{\psi}\Pi_\alpha^-[f] \cdot i_1].$$

Then

$$f = ({}^{\psi}\mathcal{K}_0 + M^\alpha \cdot {}^{\psi}V_0)[{}^{\psi}\Pi^+[f]] - (({}^{\psi}\mathcal{K}_0 - M^\alpha\, {}^{\psi}V_0)[{}^{\psi}\Pi^-[f]i_1])i_1 =$$

$$= ({}^{\psi}\mathcal{K}_0 + M^\alpha \cdot {}^{\psi}V_0)(-(2\alpha_1)^{-1}\, {}^{\psi}\mathcal{D}_\alpha M^{i_1})[f] - M^{i_1}({}^{\psi}\mathcal{K}_0 - M^\alpha \cdot {}^{\psi}V_0)[-(2\alpha_1)^{-1}\, {}^{\psi}\mathcal{D}_{-\alpha}[f]] =$$

$$= -(2\alpha_1)^{-1}({}^{\psi}\mathcal{D}[f]i_1 + {}^{\psi}K_0[f]i_1\alpha + {}^{\psi}V_0\, {}^{\psi}\mathcal{D}[f]i_1\alpha - {}^{\psi}D[f]i_1 + {}^{\psi}K_0[f]\alpha i_1 + {}^{\psi}V_0\, {}^{\psi}\mathcal{D}[f]\alpha i_1) =$$

$$= {}^{\psi}K_0[f] + {}^{\psi}V_0\, {}^{\psi}\mathcal{D}[f].$$

□

Using the representation (4.31) we prove the following

**4.21 Theorem** *Let* $\lambda \neq 0$, $\lambda \in \mathbb{H}(\mathbb{C})$, $\alpha^2 = \lambda$, $f \in C^2(\Omega) \cap C^1(\bar{\Omega})$. *Then: if* $\lambda \notin \mathfrak{S}$, *then* $f \in \ker\Delta_\lambda$ *iff*

$$f = {}^{\psi}K_\alpha[{}^{\psi}\Pi_\alpha[f]] + {}^{\psi}K_{-\alpha}[{}^{\psi}\Pi_{-\alpha}[f]]; \tag{4.32}$$

*if* $\lambda \in \mathfrak{S}$, *then* $f \in \ker\Delta_\lambda$ *iff*

$$f = {}^\psi K_{2\alpha_0}[{}^\psi\Pi_\alpha[f]] + {}^\psi K_{-2\alpha_0}[{}^\psi\Pi_{-\alpha}[f]] + \frac{1}{2\alpha_0}\,{}^\psi\mathcal{K}_0[f]\bar{\alpha} + \frac{1}{2\alpha_0}\,{}^\psi V_0\,{}^\psi\mathcal{D}[f]\bar{\alpha}, \qquad (4.33)$$

*where the projection operators* ${}^\psi\Pi_{\pm\alpha}$ *are defined in Subsection 2.8.*

PROOF. First we prove (4.32). From Proposition 2.9 and Theorem 2.13 we have (if $\lambda \notin \mathfrak{S}$)

$$f \in \ker\Delta_\lambda \iff f = {}^\psi\Pi_\alpha[f] + {}^\psi\Pi_{-\alpha}[f],$$

where ${}^\psi\Pi_\alpha[f] \in \ker{}^\psi\mathcal{D}_\alpha$, ${}^\psi\Pi_{-\alpha}[f] \in \ker{}^\psi\mathcal{D}_{-\alpha}$. Further, ${}^\psi\Pi_{\pm\alpha}[f] = {}^\psi K_{\pm\alpha}[{}^\psi\Pi_{\pm\alpha}[f]]$ from which we obtain (4.32).

Let $\lambda \in \mathfrak{S}$. Then from Theorem 2.13,

$$f \in \ker\Delta_\lambda \iff f = g + h + u,$$

where $g \in M^\alpha(\ker{}^\psi\mathcal{D}_{2\alpha_0})$; $h \in M^\alpha(\ker({}^\psi\mathcal{D}_{-2\alpha_0}))$; $u \in M^{\bar{\alpha}}(\ker\Delta)$ and $u = \frac{1}{2\alpha_0} f\bar{\alpha}$. Further, im ${}^\psi\Pi_{\pm\alpha} = M^\alpha(\ker{}^\psi\mathcal{D}_{\pm 2\alpha_0})$. Hence $g = {}^\psi\Pi_\alpha[f]$; $h = {}^\psi\Pi_{-\alpha}[f]$. Using the quaternionic Cauchy integral formula (Theorem 4.12) for the functions $g, h$ and using Theorem 4.20 for the function $u$ we obtain (4.33). □

# 5 Boundary value properties of $\alpha$-holomorphic functions

**5.1** A study of integral representations for various functional classes, which is the main theme of the book, is closely related to the study of the behaviour of corresponding functions near the boundary.

We discuss this now. As can be expected (because of the analogy both with one-dimensional complex analysis and with harmonic quaternionic analysis), the main results are conveniently expressed in terms of the operators ${}^\psi\mathcal{S}_\alpha, {}^\psi P_\alpha, {}^\psi\mathcal{Q}_\alpha$.

**5.2** Let, as above, $\Gamma$ be a closed Liapunov surface in $\mathbb{R}^3$ which is the boundary of a bounded domain $\Omega =: \Omega^+$ and of an unbounded domain $\Omega^- := \mathbb{R}^3 \setminus \overline{\Omega^+}$. Let $f : \Gamma \to \mathbb{H}(\mathbb{C})$ be an integrable function (in the Riemann sense). Then by formulas (4.21) we define the function

$$^{\psi}K_\alpha[f] : x \in \mathbb{R}^3 \setminus \Gamma \mapsto {}^{\psi}K_\alpha[f](x).$$

This will be called the (left) $(\psi, \alpha)$– holomorphic Cauchy–type integral (c.t.i.) of the function $f$. The corresponding map

$$^{\psi}K_\alpha : f \mapsto {}^{\psi}K_\alpha[f]$$

has the natural name c.t.i. It is clear that for any integrable $f$,

$$^{\psi}K_\alpha[f] \in C^\infty(\mathbb{R}^3 \setminus \Gamma; \mathbb{H}(\mathbb{C})) \cap {}^{\psi}\mathfrak{M}_\alpha(\mathbb{R}^3 \setminus \Gamma; \mathbb{H}(\mathbb{C})).$$

If $f$ is the restriction to $\Gamma$ of a function from ${}^{\psi}\mathfrak{M}_\alpha(\Omega) \cap C(\bar{\Omega})$ then, of course, the c.t.i. of $f$ equals its Cauchy integral, see Subsection 4.12.

**5.3 Theorem** (Plemelj–Sokhotski formulas for $(\psi, \alpha)$-holomorphic functions) *Let $\Gamma$ be a closed Liapunov surface, $f \in C^{0,\epsilon}(\Gamma, \mathbb{H}(\mathbb{C}))$, $0 < \epsilon \leq 1$. Then everywhere on $\Gamma$ the following limits exist:*

$$\lim_{\Omega^\pm \ni x \to \tau \in \Gamma} {}^{\psi}K_\alpha[f](x) =: {}^{\psi}K_\alpha[f]^\pm(\tau), \tag{5.1}$$

*and the following formulas hold*

$$^{\psi}K_\alpha[f]^+(\tau) = {}^{\psi}P_\alpha[f](\tau), \quad {}^{\psi}K_\alpha[f]^-(\tau) = -{}^{\psi}Q_\alpha[f](\tau), \tag{5.2}$$

*where the operators ${}^{\psi}P_\alpha$, ${}^{\psi}Q_\alpha$ are introduced in Subsection 4.10 and 4.15 and the integrals in (5.2) exist in the sense of the Cauchy principal value.*

PROOF. Again this consists naturally of two essential cases: $\alpha$ a complex number and $\alpha$ an arbitrary complex quaternion. Let first $\alpha = \alpha_0 \in \mathbb{C}$. We have

$$\mathcal{K}_{\psi,\alpha_0}(x) = -\alpha_0(4\pi|x|)^{-1}e^{-i\alpha_0|x|} +$$

$$+\mathcal{K}_{\psi}(x)\cdot(e^{-i\alpha_0|x|} + i\alpha_0|x|e^{-i\alpha_0|x|}).$$

Having expanded the function $e^{-i\alpha_0|x|}$ into the Taylor series we get:

$$^{\psi}K_{\alpha_0}[f](x) = -\int_{\Gamma}\{-\frac{\alpha_0 e^{-i\alpha_0|x-y|}}{4\pi|x-y|} + i\alpha_0\mathcal{K}_{\psi}(x-y)|x-y|e^{-i\alpha_0|x-y|} +$$

$$+\mathcal{K}_{\psi}(x-y)\sum_{k=1}^{\infty}\frac{(-i\alpha_0|x-y|)^k}{k!}\}\vec{n}_{\psi}(y)f(y)d\Gamma_y -$$

$$-\int_{\Gamma}\mathcal{K}_{\psi}(x-y)\vec{n}_{\psi}(y)f(y)d\Gamma_y. \tag{5.3}$$

For the last term (which is nothing more than the c.t.i. $^{\psi}K[f]$ for $\alpha = 0$) all assertions of Theorem 5.3 are well-known and the proof can be found, for instance, in [14, p. 177]; [46, p. 59]; [111]. Thus we have that the limit $\lim_{\Omega^{\pm}\ni x\to\tau\in\Gamma} {^{\psi}K}[f](x)$ exists and

$$\lim_{\Omega^{+}\ni x\to\tau\in\Gamma} {^{\psi}K}[f](x) = {^{\psi}P}[f](\tau), \quad \lim_{\Omega^{-}\ni x\to\tau\in\Gamma} {^{\psi}K}[f](x) = -{^{\psi}Q}[f](\tau). \tag{5.4}$$

Thus the case $\alpha = \alpha_0 \in \mathbb{C}$ is covered. This becomes immediately the base for the proof of the general case $\alpha \in \mathbb{H}(\mathbb{C})$ when we take into account the defining formulas (4.21). These give without any calculations the existence of the limits (5.1). Formulas (5.2) do require some simple calculations. For instance, let $\alpha \notin \mathfrak{S}$, $\vec{\alpha}^2 = 0$. Then

$$^{\psi}K_{\alpha}[f]^{+}(\tau) = {^{\psi}P_{\alpha_0}}[f](\tau) + M^{\vec{\alpha}}\frac{\partial}{\partial\alpha_0}{^{\psi}P_{\alpha_0}}[f](\tau) =$$

$$= \frac{1}{2}(f(\tau) + {^{\psi}S_{\alpha_0}}[f](\tau)) + \frac{1}{2}M^{\vec{\alpha}}\frac{\partial}{\partial\alpha_0}{^{\psi}S_{\alpha_0}}[f](\tau) =$$

$$= \frac{1}{2}f(\tau) + \frac{1}{2}{^{\psi}S_{\alpha}}[f](\tau) =: {^{\psi}P_{\alpha}}[f](\tau).$$

In the case $\alpha \in \mathfrak{S}$, $\alpha_0 = 0$ we must take into account the continuity of the single-layer potential $V_0[f]$, which gives

$$\begin{aligned}
{}^{\psi}\mathcal{K}_\alpha[f]^+(\tau) &= {}^{\psi}P_0[f](\tau) - M^\alpha \cdot {}^{\psi}V_0[f](\tau) = \\
&= \frac{1}{2}(f(\tau) + {}^{\psi}S[f](\tau) - M^\alpha \cdot {}^{\psi}\hat{V}_0[f](\tau)) = \\
&= \frac{1}{2}(f(\tau) + {}^{\psi}S_\alpha[f](\tau)) =: {}^{\psi}P_\alpha[f](\tau).
\end{aligned}$$

□

**5.4 Observation** In Subsection 5.3, $f$ is continuous and hence all integrals are understood in the Riemann sense (proper or improper). Let now $f \in L_p(\Gamma; \mathbb{H}(\mathbb{C}))$, $p > 1$. Then one has to understand ${}^{\psi}\mathcal{K}_\alpha[f]$ as a Lebesgue integral, and the necessary changes can be easily made. For instance, the limits (5.1) exist only almost everywhere on $\Gamma$ (with respect to the surface Lebesgue measure), and the formulas (5.2) are valid almost everywhere as well. The proofs use some standard tricks, see, e.g., the proof of Theorem 5.5.

In what follows, we shall use the $L_p$-formulation of Theorem 5.3 if necessary, without any special remark.

**5.5 Theorem** (Involutiveness of the operator ${}^{\psi}\mathcal{S}_\alpha$) *The operator ${}^{\psi}\mathcal{S}_\alpha$ is an involution on the complex spaces $C^{0,\epsilon}(\Gamma, \mathbb{H}(\mathbb{C}))$, $0 < \epsilon \leq 1$, and $L_p(\Gamma, \mathbb{H}(\mathbb{C}))$, $p > 1$, and hence ${}^{\psi}\mathcal{P}_\alpha$ and ${}^{\psi}\mathcal{Q}_\alpha$ are mutually complementary projectors on these functional spaces:*

$$ {}^{\psi}\mathcal{S}_\alpha^2 = I, \tag{5.5}$$

$$ {}^{\psi}\mathcal{P}_\alpha^2 = {}^{\psi}P_\alpha, \quad {}^{\psi}\mathcal{Q}_\alpha^2 = {}^{\psi}Q_\alpha; \qquad {}^{\psi}\mathcal{P}_\alpha \cdot {}^{\psi}Q_\alpha = {}^{\psi}Q_\alpha \cdot {}^{\psi}P_\alpha = 0. \tag{5.6}$$

PROOF. (5.5) and (5.6) are equivalent, so it is enough to prove (5.6). Let $f \in C^{0,\epsilon}(\Gamma; \mathbb{H}(\mathbb{C}))$, then ${}^{\psi}\mathcal{K}_\alpha[f] \in {}^{\psi}\mathfrak{M}_\alpha(\Omega) \cap C(\bar{\Omega})$ and we can apply the quaternionic Cauchy integral formula to the function ${}^{\psi}\mathcal{K}_\alpha[f] : \forall x \in \Omega^+$ :

$${}^{\psi}\!\mathcal{K}_\alpha[f](x) = {}^{\psi}K_\alpha[{}^{\psi}\!\mathcal{K}_\alpha[f]](x).$$

Now letting $x \to \tau \in \Gamma$ and using (5.2) we obtain the necessary result.

Assuming that (5.5) is true on $C^{0,\epsilon}$ we obtain immediately that (5.5) is true on $L_p, p > 1$, recalling that $C^{0,\epsilon}$ is dense in $L_p$ and that $I$ is a continuous operator. □

**5.6** It is interesting to mention the following. The equality (5.5) holds for an arbitrary $\alpha$, in particular, for $\alpha = 0$ we have ${}^{\psi}\!\mathcal{S}^2 = I$. On the other hand, using the defining equalities (4.22) from Subsection 4.15 we will prove in Subsection 6.8 that the difference ${}^{\psi}\!\mathcal{S}_\alpha - {}^{\psi}S =: {}^{\psi}\!A_\alpha$ is a compact operator (its kernel has a weak singularity). Hence, for the involutive operator ${}^{\psi}\!\mathcal{S}$ we have constructed a family of compact operators $\{{}^{\psi}\!A_\alpha | \alpha \in \mathbb{H}(\mathbb{C})\}$ such that an additive perturbation of ${}^{\psi}\!\mathcal{S}$ by elements of the family does not keep it from being involutive.

Very similar statements can be made about ${}^{\psi}\!\mathcal{P}_\alpha$ and ${}^{\psi}\!\mathcal{Q}_\alpha$: again we have projectors ${}^{\psi}\!\mathcal{P}$ and ${}^{\psi}\!\mathcal{Q}$ and perturbing them additively by some compact operators we obtain projectors as well. From Theorem 5.5 it follows that both ${}^{\psi}\!\mathcal{P}_\alpha \cdot {}^{\psi}Q_\alpha$ and ${}^{\psi}\!\mathcal{Q}_\alpha \cdot {}^{\psi}P_\alpha$ are zero operators on the corresponding spaces.

Moreover, the following assertions show that ${}^{\psi}\!\mathcal{P}_\alpha$ is the operator which projects each of the spaces $C^{0,\epsilon}(\Gamma)$ and $L_p(\Gamma)$ onto their respective subsets of all functions $(\psi, \alpha)$-holomorphically extendable from the boundary into $\Omega^+$. Analogously, ${}^{\psi}\!\mathcal{Q}_\alpha$ projects these spaces onto the functions $(\psi, \alpha)$-holomorphically extendable into $\Omega^-$.

**5.7 Theorem** (**$(\psi, \alpha)$-holomorphic extension of a given Hölder function**) *Let $\Gamma$ be a closed Liapunov surface which is the boundary of a finite domain $\Omega^+$ and of an infinite domain $\Omega^-$. Let $f \in C^{0,\epsilon}(\Gamma, \mathbb{H}(\mathbb{C}))$, $0 < \epsilon \leq 1$.*

**1)** *In order for $f$ to be a boundary value (i.e., a trace on $\Gamma$) of a function $\tilde{f}$ from ${}^{\psi}\mathfrak{M}_\alpha(\Omega^+) \cap C^{0,\epsilon}(\overline{\Omega^+})$, the following condition is necessary and sufficient:*

$$f(\tau) = {}^{\psi}S_\alpha[f](\tau), \quad \forall \tau \in \Gamma, \tag{5.7}$$

*or, equivalently,*

$$f \in \mathrm{im}\,{}^{\psi}\mathcal{P}_\alpha := {}^{\psi}P_\alpha(C^{0,\epsilon}). \tag{5.8}$$

2) *In order for $f$ to be a boundary value of a function $\tilde{f}$ from ${}^{\psi}\mathfrak{M}_\alpha(\Omega^-) \cap C^{0,\epsilon}(\bar{\Omega}^-)$, $\tilde{f}(\infty) = 0$, the following condition is necessary and sufficient:*

$$f(\tau) = -{}^{\psi}\mathcal{S}_\alpha[f](\tau), \quad \forall \tau \in \Gamma, \tag{5.9}$$

*or, equivalently,*

$$f \in \mathrm{im}\,{}^{\psi}\mathcal{Q}_\alpha := {}^{\psi}Q_\alpha(C^{0,\epsilon}). \tag{5.10}$$

PROOF. First let $f \in C^{0,\epsilon}(\Gamma)$ be the boundary value of $\tilde{f} \in {}^{\psi}\mathfrak{M}(\Omega^+) \cap C^{0,\epsilon}(\overline{\Omega^+})$. Then $\tilde{f}$ is representable by its Cauchy integral:

$$\tilde{f}(x) = {}^{\psi}K_\alpha[f](x), \quad \forall x \in \Omega^+. \tag{5.11}$$

Let now $\tau \in \Gamma$ and $\Omega^+ \ni x \to \tau$. By the conditions of the theorem and by the Plemelj–Sokhotski formulas we obtain

$$f(\tau) = {}^{\psi}P_\alpha[f](\tau) = \frac{1}{2}(f(\tau) + {}^{\psi}S_\alpha[f](\tau))$$

which gives both (5.8) and (5.7).

Now, on the contrary, let (5.8) hold. Denote ${}^{\psi}\mathcal{K}_\alpha[f](x) =: \tilde{f}(x)$ $\forall x \in \Omega^+$. This gives immediately that $\tilde{f} \in {}^{\psi}\mathfrak{M}_\alpha(\Omega^+) \cap C^{0,\epsilon}(\overline{\Omega^+})$. Further, by the Plemelj-Sokhotski formulas, $\tilde{f}|_\Gamma = {}^{\psi}P_\alpha[f] = f$ (${}^{\psi}\mathcal{P}_\alpha$ is a projector).

The first part is completely proved. The proof of the second part is quite analogous.

□

5.8 To formulate corresponding assertions for $f \in L_p(\Gamma, \mathbb{H}(\mathbb{C}))$, the following definition is useful. If $X$ is any space consisting of $\mathbb{H}(\mathbb{C})$-valued functions defined on $\Gamma$, then ${}^{\psi}\mathfrak{A}_\alpha(\Omega^\pm; X)$ denotes classes of functions $\tilde{f}^\pm$ such that

1) $\tilde{f}^\pm \in {}^{\psi}\mathfrak{M}_\alpha(\Omega^\pm)$;

2) for $\Omega^\pm \ni x \to \tau \in \Gamma$ there exist $\lim \tilde{f}^\pm(x) =: f^\pm(\tau)$ (at any rate, along non-tangential paths) everywhere or almost everywhere on $\Gamma$ generating the function $f^\pm$ from $X$;

3) $\tilde{f}^\pm$ is representable by the left-$(\psi, \alpha)$-holomorphic c.t.i. with a density from $X$.

If $X = L_p(\Gamma; \mathbb{H}(\mathbb{C}))$, then we write ${}^{\psi}\mathfrak{A}_{\alpha,p}(\Omega^\pm)$. It is clear that $L_p(\Gamma; \mathbb{H}(\mathbb{C}))$ is a Banach $\mathbb{H}(\mathbb{C})$-bimodule as well as a complex space. The classes ${}^{\psi}\mathfrak{A}_{\alpha,p}(\Omega^\pm)$ generate subspaces of $L_p(\Gamma, \mathbb{H}(\mathbb{C}))$. One can show that for $\Omega^+ = \mathbb{B}(0;1) := \{x | x_1^2 + x_2^2 + x_3^2 < 1\}$ the definition of the class ${}^{\psi}\mathfrak{A}_{\alpha,p}(\Omega^+)$ is a natural $(\psi, \alpha)$-holomorphic analogue of the definition of the complex Hardy space $H_p(\mathbb{B})$. Therefore we shall use the notation ${}^{\psi}\mathfrak{H}_{\alpha,p}(\mathbb{B})$ for ${}^{\psi}\mathfrak{A}_{\alpha,p}(\mathbb{B})$. As usual we identify an element of ${}^{\psi}\mathfrak{H}_{\alpha,p}(\mathbb{B}, \mathbb{H}(\mathbb{C}))$ with the corresponding function from $L_p(\Gamma; \mathbb{H}(\mathbb{C}))$ generated by its limit values. Just in this sense ${}^{\psi}\mathfrak{H}_{\alpha,p}(\mathbb{B}; \mathbb{H}(\mathbb{C}))$ is a subspace of $L_p(\Gamma; \mathbb{H}(\mathbb{C}))$.

**5.9 Theorem** ($(\psi, \alpha)$-holomorphic extension of a given $L_p$-function) *Let $\Gamma$ be a closed Liapunov surface which is the boundary of a finite domain $\Omega^+$ and of an infinite domain $\Omega^-$. Let $f \in L_p(\Gamma; \mathbb{H}(\mathbb{C}))$.*

1) *In order for the values of the function $f$ to coincide almost everywhere on $\Gamma$ with the limit values of a function $\tilde{f} \in {}^{\psi}\mathfrak{A}_{\alpha,p}(\Omega^+)$, the following condition is necessary and sufficient:*

$$f(\tau) = {}^{\psi}S_\alpha[f](\tau)$$

*almost everywhere on $\Gamma$, or, equivalently,*

$$f \in \operatorname{im}{}^{\psi}\mathcal{P}_\alpha := {}^{\psi}P_\alpha(L_p(\Gamma)).$$

2) *In order for the values of the function* $f$ *to coincide almost everywhere on* $\Gamma$ *with the limit values of a function* $\tilde{f} \in {}^{\psi}\mathfrak{A}_{\alpha,p}(\Omega^-)$, *the following condition is necessary and sufficient:*

$$f(\tau) = -{}^{\psi}\mathcal{S}_\alpha[f](\tau)$$

*almost everywhere on* $\Gamma$, *or, equivalently,*

$$f \in \operatorname{im}{}^{\psi}\mathcal{Q}_\alpha(L_p(\Gamma)).$$

**5.10** Remark Till now we assumed that $\Gamma$ was a finite surface. But also we shall need to consider infinite surfaces, most often hyperplanes. Having analysed the proofs of this section one can see that they remain true provided that the integrals involved are convergent at infinity. For instance, the proof of the Plemelj–Sokhotski formulas for ${}^{\psi}K_\alpha[f]$ is essentially local, and is true both for Liapunov non-closed surfaces (except for points of the boundary of this surface) and for hyperplanes.

**5.11** We now establish a very good analogue, for $\alpha$-holomorphic functions, of the so-called Hilbert formulas from one-dimensional complex analysis. With this in mind, we shall describe here briefly what occurs in the complex situation.

Let $\mathbb{C}_+$ denote the upper half-plane of the complex plane $\mathbb{C}$ and let $\Gamma = \mathbb{R}$ denote its boundary, the real axis in $\mathbb{C}$. Let $\tilde{f}$ be holomorphic (in the usual, complex one-dimensional sense) in $\mathbb{C}_+$ and have the limit function $f$ on $\mathbb{R}$ of the class $L_p(\mathbb{R})$, $p > 1$.

We are looking for how to express one of the real components of $f =: f_1 + i f_2$ via the other. Or, which is equivalent, how to construct the limit function $f$ knowing one of its real components. Of course, this has one more interpretation: since having the limit function $f$ we have the function $\tilde{f}$ also; then we are looking, in fact, for $\tilde{f}$ provided we know one of the real components of its limit function $f$.

How may one solve this problem? Let $S_{\mathbb{R}}$ denote the singular integration operator with the complex Cauchy kernel along $\mathbb{R}$:

$$S_{\mathbb{R}}[f](\tau) := \frac{1}{\pi i}\int_{\mathbb{R}} \frac{f(x)}{x-\tau}dx, \quad \tau \in \mathbb{R}. \tag{5.12}$$

It is known that $S_{\mathbb{R}}$ is an involution on $L_p(\mathbb{R},\mathbb{C}) : S_{\mathbb{R}}^2 = I$, the identity operator. Also, a necessary and sufficient condition for $f \in L_p(\mathbb{R})$ to be the limit value of a holomorphic function $\tilde{f}$ in $\mathbb{C}_+$ is that

$$f = S_{\mathbb{R}}[f] \tag{5.13}$$

almost everywhere on $\mathbb{R}$.

For our purposes it is important to note that the kernel $\frac{1}{\pi i}\frac{1}{x-\tau}$ is purely imaginary. We have from (5.13):

$$f_1 + if_2 = S_{\mathbb{R}}[f_1 + if_2] = S_{\mathbb{R}}[f_1] + iS_{\mathbb{R}}[f_2],$$

which yields

$$f_1(\tau) = \frac{1}{\pi}\int_{\mathbb{R}} \frac{f_2(x)}{x-\tau}dx, \tag{5.14}$$

$$f_2(\tau) = -\frac{1}{\pi}\int_{\mathbb{R}} \frac{f_1(x)}{x-\tau}dx. \tag{5.15}$$

These two formulas are often called the mutually inverse Hilbert formulas. They solve the above-mentioned problems. For instance: for $\forall z \in \mathbb{C}_+$,

$$\begin{aligned} \tilde{f}(z) &:= K[f_1 + S_{\mathbb{R}}[f_1]](z) = \\ &= K[iS_{\mathbb{R}}[f_2] + f_2](z), \end{aligned} \tag{5.16}$$

where $K$ is the usual complex Cauchy integral.

The operator in (5.14) relating $\mathrm{Re}f$ and $\mathrm{Im}f$ has a special name: the Hilbert operator or the Hilbert transform. Thus

$$H_{\mathbb{R}}[f](\tau) := \frac{1}{\pi}\int_{\mathbb{R}} \frac{f(x)}{x-\tau}dx. \tag{5.17}$$

Its kernel $\dfrac{1}{\pi}\dfrac{1}{x-\tau}$ is real-valued and hence $H_{\mathbb{R}}$ transforms a real-valued function into a real-valued function.

We have obviously that

$$S_{\mathbb{R}} = -iH_{\mathbb{R}}. \tag{5.18}$$

Hence

$$H_{\mathbb{R}}^2 = -I, \tag{5.19}$$

which explains in a certain sense formulas (5.14)-(5.15):

$$f_1 = H_{\mathbb{R}}[f_2], \tag{5.20}$$

$$f_2 = -H_{\mathbb{R}}[f_1]. \tag{5.21}$$

The reciprocity of the formulas (5.20) and (5.21) is a corollary of the invertibility of $H_{\mathbb{R}} : H_{\mathbb{R}}^{-1} = -H_{\mathbb{R}}$.

**5.12** Developing these ideas to the hyperholomorphic situation we shall consider, for the sake of simplicity and taking into account forthcoming applications, the case $\psi = \psi_{st}$ only. Firstly we examine the case $\alpha_0 = 0$. Then the operator $S_\alpha$ has the same structure both for $\alpha = \vec{\alpha} \in \mathfrak{S}$ and $\alpha = \vec{\alpha} \notin \mathfrak{S}$ given by (4.23).

We assume now that

$$\Gamma := \mathbb{R}^2 := \{x \in \mathbb{R}^3 | x_3 = 0\}, \vec{n} = -i_3. \tag{5.22}$$

Introduce the notation:

$$R_{\gamma,k}[\nu](x) := 2\int_{\mathbb{R}^2}\theta_\gamma(x-y)\left(\frac{x_k-y_k}{|x-y|^2}+i\gamma\frac{x_k-y_k}{|x-y|}\right)\nu(y)dy \tag{5.23}$$

where $k \in \mathbb{N}_2$; $\gamma \in \mathbb{C}$, $\gamma^2 = \vec{\alpha}^2$; $\theta_\gamma$ is defined in Subsection 2.19. The integrals (5.23) exist, for instance, if $\nu \in L_p(\mathbb{R}^2;\mathbb{C})$ (see [81]). We shall call them the acoustic, or metaharmonic, Riesz transforms because in the limit case $\gamma = 0$ we get the famous Riesz transforms $R_k := R_{0,k}$ in $\mathbb{R}^2$. For an arbitrary $m \geq 1$ in $\mathbb{R}^m$ there exist $m$ Riesz transforms $R_k$, $k \in \mathbb{N}_m$, and for $m = 1$ the only Riesz transform $R_1$ in $\mathbb{R}$ coincides with the Hilbert transform $H_\mathbb{R}$ from (5.17). That is why for $m > 1$ each $R_k$ from the collection $\{R_1, \ldots, R_m\}$ is considered to be a good multidimensional analogue of $H_\mathbb{R}$. We would like to emphasize here that the formula (5.18) hints at a way to introduce the unique multidimensional generalization of $H_\mathbb{R}$.

Let us mention also that the Riesz transform possesses the following important properties:

$$\sum_k R_k^2 = -I, \tag{5.24}$$

$$R_k R_q = R_q R_k, \quad \forall k, q, \tag{5.25}$$

see [121, p. 224], for instance.

Returning to $R_{\gamma,k}$ let us use (4.23) to express $S_\alpha$ via $R_{\gamma,k}$:

$$S_\alpha = -R_{\gamma,1}i_2 + R_{\gamma,2}i_1 - M^{\vec{\alpha}} \cdot i_3\check{V}_\gamma \tag{5.26}$$

where $\check{V}_\gamma[f](x) = 2\int_{\mathbb{R}^2}\theta_\gamma(x-\tau)f(\tau)d\tau$. It is essential that all three operators $R_{\gamma,1}$; $R_{\gamma,2}$; $\check{V}_\gamma$ have complex-valued kernels. The equality (5.5) allows us to arrive at the following generalizations of (5.24) and (5.25).

**5.13 Proposition** (Properties of the acoustic Riesz transforms) *Let $\gamma$ be an arbitrary complex number, $R_{\gamma,k}$ be from (5.23). Then the following operator equalities are valid* *on* $L_p(\Gamma;\mathbb{C})$ *:*

$$R^2_{\gamma,1}+R^2_{\gamma,2}+\gamma^2\check{V}_\gamma=-I; \tag{5.27}$$

$$R_{\gamma,k}\check{V}_\gamma=\check{V}_\gamma R_{\gamma,k}, \quad k\in\mathbb{N}_2; \tag{5.28}$$

$$R_{\gamma,1}R_{\gamma,2}=R_{\gamma,2}R_{\gamma,1}. \tag{5.29}$$

PROOF. Define $\alpha_1=\alpha_2=0$, $\alpha_3=i\gamma$, then $\alpha:=\sum_{k=1}^3\alpha_k i_k$ satisfies the condition $\alpha^2=\gamma^2$. Let $S_\alpha$ denote the corresponding operator (5.26). Putting (5.26) into (5.5) and separating coordinates in the "quaternionic" equality (5.5) we arrive at the system of four "complex" equalities giving (5.27)–(5.29). □

**5.14** Let again $\alpha=\vec{\alpha}$, $S_\alpha$ be defined by (5.26). Consider the equation

$$S_\alpha[f]=g \tag{5.30}$$

with $g\in L_p(\mathbb{R}^2;\mathbb{H}(\mathbb{C}))$. Its unique solution in the same space is given by

$$f:=S_\alpha[g] \tag{5.31}$$

because of (5.5). Writing as usual $f=\sum_{k=0}^3 f_k i_k$, $g=\sum_{q=0}^3 g_q i_q$, where $\{f_k,g_q|\forall k,q\}$ are complex-valued functions, we get from (5.30) and (5.26):

$$(\alpha_3\check{V}_\gamma+i_1(R_{\gamma,2}+\alpha_2\check{V}_\gamma)-i_2(R_{\gamma,1}+\alpha_1\check{V}_\gamma))[\sum_{k=0}^3 f_k i_k]=\sum_{q=0}^3 g_q i_q,$$

or in coordinate-wise form

$$
\begin{cases}
-(R_{\gamma,2}+\alpha_2\check{V}_\gamma)[f_1]+(R_{\gamma,1}+\alpha_1\check{V}_\gamma)[f_2]+\alpha_3\check{V}_\gamma[f_0]=g_0,\\
(R_{\gamma,2}+\alpha_2\check{V}_\gamma)[f_0]-(R_{\gamma,1}+\alpha_1\check{V}_\gamma)[f_3]+\alpha_3\check{V}_\gamma[f_1]=g_1,\\
-(R_{\gamma,2}+\alpha_2\check{V}_\gamma)[f_3]-(R_{\gamma,1}+\alpha_1\check{V}_\gamma)[f_0]+\alpha_3\check{V}_\gamma[f_2]=g_2,\\
(R_{\gamma,2}+\alpha_2\check{V}_\gamma)[f_2]+(R_{\gamma,1}+\alpha_1\check{V}_\gamma)[f_1]+\alpha_3\check{V}_\gamma[f_3]=g_3.
\end{cases}
\tag{5.32}
$$

The same holds for (5.31):

$$
\begin{cases}
-(R_{\gamma,2}+\alpha_2\check{V}_\gamma)[g_1]+(R_{\gamma,1}+\alpha_1\check{V}_\gamma)[g_2]+\alpha_3\check{V}_\gamma[g_0]=f_0,\\
(R_{\gamma,2}+\alpha_2\check{V}_\gamma)[g_0]-(R_{\gamma,1}+\alpha_1\check{V}_\gamma)[g_3]+\alpha_3\check{V}_\gamma[g_1]=f_1,\\
-(R_{\gamma,2}+\alpha_2\check{V}_\gamma)[g_3]-(R_{\gamma,1}+\alpha_1\check{V}_\gamma)[g_0]+\alpha_3\check{V}_\gamma[g_2]=f_2,\\
(R_{\gamma,2}+\alpha_2\check{V}_\gamma)[g_2]+(R_{\gamma,1}+\alpha_1\check{V}_\gamma)[g_1]+\alpha_3\check{V}_\gamma[g_3]=f_3.
\end{cases}
\tag{5.33}
$$

Assume now that $\alpha_3 = 0$. It is easily seen that each of the systems (5.32) and (5.33) separates into two independent systems: for (5.32) we have

$$
\begin{cases}
-(R_{\gamma,2}+\alpha_2\check{V}_\gamma)[f_1]+(R_{\gamma,1}+\alpha_1\check{V}_\gamma)[f_2]=g_0,\\
(R_{\gamma,1}+\alpha_1\check{V}_\gamma)[f_1]+(R_{\gamma,2}+\alpha_2\check{V}_\gamma)[f_2]=g_3,
\end{cases}
\tag{5.34}
$$

and

$$
\begin{cases}
(R_{\gamma,2}+\alpha_2\check{V}_\gamma)[f_0]-(R_{\gamma,1}+\alpha_1\check{V}_\gamma)[f_3]=g_1,\\
-(R_{\gamma,1}+\alpha_1\check{V}_\gamma)[f_0]-(R_{\gamma,2}+\alpha_2\check{V}_\gamma)[f_3]=g_2,
\end{cases}
\tag{5.35}
$$

for (5.33) analogously

$$\begin{cases} -(R_{\gamma,2}+\alpha_2\check{V}_\gamma)[g_1]+(R_{\gamma,1}+\alpha_1\check{V}_\gamma)[g_2]=f_0, \\ \\ (R_{\gamma,1}+\alpha_1\check{V}_\gamma)[g_1]+(R_{\gamma,2}+\alpha_2\check{V}_\gamma)[g_2]=f_3, \end{cases} \tag{5.36}$$

and

$$\begin{cases} (R_{\gamma,2}+\alpha_2\check{V}_\gamma)[g_0]-(R_{\gamma,1}+\alpha_1\check{V}_\gamma)[g_3]=f_1, \\ \\ -(R_{\gamma,1}+\alpha_1\check{V}_\gamma)[g_0]-(R_{\gamma,2}+\alpha_2\check{V}_\gamma)[g_3]=f_2. \end{cases} \tag{5.37}$$

Thus, in fact we have the following. For an arbitrary $\alpha_3$ the system (5.33) determines uniquely the solution of (5.32), and vice versa. For $\alpha_3 = 0$ the unknown functions become separated in such a way that the pair $(f_1, f_2)$ takes part in the system (5.34) only, and the pair $(f_0, f_3)$ in (5.35) only. Moreover, both systems coincide, hence it is enough to be able to solve one of them only.

By construction, (5.37) gives the solution to (5.34), and (5.36) gives the solution to (5.35).

**5.15** It appears that the conclusions made above as well as the forthcoming Hilbert formulas become more transparent and comprehensible if we use one more algebraic structure: that of bicomplex numbers. Let $f \in \mathbb{H}(\mathbb{C})$, then

$$f = \sum_{k=0}^{3} f_k i_k = f_0 + f_3 i_3 + (f_1 + f_2 i_3) i_1 =: F_1 + F_2 i_1.$$

Each complex quaternion $F_1, F_2$ is of the form $a + bi_3$ with $a, b$ usual complex numbers, and thus belongs to the algebra of bicomplex numbers generated by the imaginary units $i$ and $i_3$ which we shall denote by $\mathbb{C}(i) \otimes \mathbb{C}(i_3)$. $F_1$ and $F_2$ will be called bicomplex components (or coordinates) of $f$. Conjugation with respect to $i_3$ will be denoted as follows:

$$\hat{F}_1 := f_0 - f_3 i_3.$$

The corresponding operator will be denoted sometimes as $\hat{Z} : \hat{Z}[F_1] := \hat{F}_1$. It is clear that $F_1 i_1 = i_1 \hat{F}_1$; i.e., $M^{i_1}[F_1] = i_1 \hat{Z}[F_1]$, and also $Z_{\mathbf{H}}[F_1] = \hat{Z}[F_1]$.

In bicomplex terms we have for the operator (5.26):

$$S_\alpha[F_1 + F_2 i_1] = -(R_{\gamma,2} - R_{\gamma,1} i_3)[\hat{F}_2] - i_3 \check{V}_\gamma[\alpha_3 i_3 F_1 - (\alpha_1 - \alpha_2 i_3) F_2] +$$
$$+((R_{\gamma,2} - R_{\gamma,1} i_3)[\hat{F}_1] - i_3 \check{V}_\gamma[(\alpha_1 + \alpha_2 i_3) F_1 - \alpha_3 i_3 F_2]) i_1. \tag{5.38}$$

Thus the equation (5.30) is equivalent to the system of bicomplex equations

$$\begin{aligned} &\alpha_3 \check{V}_\gamma[F_1] + i_3(\alpha_1 - \alpha_2 i_3) \check{V}_\gamma[F_2] - (R_{\gamma,2} - R_{\gamma,1} i_3)[\hat{F}_2] = G_1, \\ &\alpha_3 i_3 \check{V}_\gamma[F_2] - i_3(\alpha_1 + \alpha_2 i_3) \check{V}_\gamma[F_1] + (R_{\gamma,2} - R_{\gamma,1} i_3)[\hat{F}_1] = G_2, \end{aligned} \tag{5.39}$$

and analogously for (5.31):

$$\begin{aligned} &\alpha_3 \check{V}_\gamma[G_1] + i_3(\alpha_1 - \alpha_2 i_3) \check{V}_\gamma[G_2] - (R_{\gamma,2} - R_{\gamma,1} i_3)[\hat{G}_2] = F_1, \\ &\alpha_3 i_3 \check{V}_\gamma[G_2] - i_3(\alpha_1 + \alpha_2 i_3) \check{V}_\gamma[G_1] + (R_{\gamma,2} - R_{\gamma,1} i_3)[\hat{G}_1] = F_2. \end{aligned} \tag{5.40}$$

Of course, (5.39) and (5.40) are the bicomplex forms of (5.32) and (5.33). If now $\alpha_3 = 0$ then they degenerate, respectively, into

$$\begin{cases} i_3(\alpha_1 - \alpha_2 i_3) \check{V}_\gamma[F_2] - (R_{\gamma,2} - R_{\gamma,1} i_3)[\hat{F}_2] = G_1, \\ -i_3(\alpha_1 + \alpha_2 i_3) \check{V}_\gamma[F_1] + (R_{\gamma,2} - R_{\gamma,1} i_3)[\hat{F}_1] = G_2, \end{cases} \tag{5.41}$$

(see (5.34) and (5.35)) and

$$\begin{cases} i_3(\alpha_1 - \alpha_2 i_3) \check{V}_\gamma[G_2] - (R_{\gamma,2} - R_{\gamma,1} i_3)[\hat{G}_2] = F_1, \\ -i_3(\alpha_1 + \alpha_2 i_3) \check{V}_\gamma[G_1] + (R_{\gamma,2} - R_{\gamma,1} i_3)[\hat{G}_1] = F_2. \end{cases} \tag{5.42}$$

We have proved the following

**5.16 Theorem** *Let* $\Gamma = \mathbb{R}^2$, $\vec{n} = -i_3, \alpha \in \mathbb{H}(\mathbb{C})$, $\alpha_0 = \alpha_3 = 0$. *Then each of the bicomplex coordinates of the unique solution* $f$ *to the equation* $S_\alpha[f] = g$ *depends on one bicomplex coordinate of* $g$ *only, the dependence being realized by the operators*

$$(\alpha_2 \check{V}_\gamma - R_{\gamma,2}\hat{Z}) + (\alpha_1 \check{V}_\gamma - R_{\gamma,1}\hat{Z})i_3$$

*and*

$$(\alpha_2 \check{V}_\gamma + R_{\gamma,2}\hat{Z}) - (\alpha_1 \check{V}_\gamma - R_{\gamma,1}\hat{Z})i_3.$$

**5.17 Corollary** *Under the conditions of Theorem 5.16, the operator*

$$(\alpha_2 \check{V}_\gamma - R_{\gamma,2}\hat{Z}) + (\alpha_1 \check{V}_\gamma - R_{\gamma,1}\hat{Z})i_3 \tag{5.43}$$

*is invertible on the complex (over* $\mathbb{C}(i)$*) space* $L_p(\mathbb{R}^2; \mathbb{C}(i) \otimes \mathbb{C}(i_3))$, *and its inverse is given by*

$$(\alpha_2 \check{V}_\gamma + R_{\gamma,2}\hat{Z}) - (\alpha_1 \check{V}_\gamma - R_{\gamma,1}\hat{Z})i_3. \tag{5.44}$$

**5.18 Definition** Let $\Gamma = \mathbb{R}^2$, $\vec{n} = -i_3$, $\alpha \in \mathbb{H}(\mathbb{C})$, $\alpha_0 = \alpha_3 = 0$.

The operator (5.43) will be called "the $\vec{\alpha}$-holomorphic Hilbert operator" for the case under consideration, and will be denoted by $H_{\vec{\alpha}}$:

$$H_{\vec{\alpha}} = (\alpha_2 \check{V}_\gamma - R_{\gamma,2}\hat{Z}) + (\alpha_1 \check{V}_\gamma - R_{\gamma,1}\hat{Z})i_3.$$

Corollary 5.17 says that $H_{\vec{\alpha}}$ is invertible, and its inverse is given by (5.44).

The reasons for this definition can be easily surmised. Moreover, they are completely, in our opinion, justified by the following theorem.

**5.19 Theorem** (Hilbert formulas for $\vec{\alpha}$-holomorphic functions with $\vec{\alpha} = \alpha_1 i_1 + \alpha_2 i_2$) *A function* $f$ *with bicomplex coordinates* $F_1$ *and* $F_2$ *is the boundary value of* $\tilde{f} \in$

$\mathfrak{A}_{\vec{\alpha},p}(\mathbf{R}_3^+)$, $\mathbf{R}_3^+ := \{x \in \mathbf{R}^3 | x_3 > 0\}$ *if, and only if,* $F_1$ *and* $F_2$ *are related by the formulas*

$$F_1 = H_{\vec{\alpha}}[F_2], \qquad F_2 = H_{\vec{\alpha}}^{-1}[F_1]. \tag{5.45}$$

PROOF. By Theorem 5.9 we have

$$f = S_{\vec{\alpha}}[f],$$

and we can apply formulas (5.41) or (5.42). □

Let us give an explicit form of the equalities (5.45)

$$F_1(x) = -\frac{1}{2\pi}\int_{\mathbf{R}^2} \frac{e^{-i\gamma|x-\tau|}}{|x-\tau|}(\alpha_2 + \alpha_1 i_3)F_2(\tau)d\tau$$

$$+\frac{1}{2\pi}\int_{\mathbf{R}^2} \frac{e^{-i\gamma|x-\tau|}}{|x-\tau|}\left(\left(\frac{x_1-\tau_1}{|x-\tau|^2} - i\gamma\frac{x_1-\tau_1}{|x-\tau|}\right)i_3 + \left(\frac{x_2-\tau_2}{|x-\tau|^2} - i\gamma\frac{x_2-\tau_2}{|x-\tau|}\right)\right)\hat{F}_2(\tau)d\tau,$$

$$F_2(x) = -\frac{1}{2\pi}\int_{\mathbf{R}^2} \frac{e^{-i\gamma|x-\tau|}}{|x-\tau|}(\alpha_2 - \alpha_1 i_3)F_1(\tau)d\tau$$

$$-\frac{1}{2\pi}\int_{\mathbf{R}^2} \frac{e^{-i\gamma|x-\tau|}}{|x-\tau|}\left(\left(\frac{x_1-\tau_1}{|x-\tau|^2} - i\gamma\frac{x_1-\tau_1}{|x-\tau|}\right)i_3 + \left(\frac{x_2-\tau_2}{|x-\tau|^2} - i\gamma\frac{x_2-\tau_2}{|x-\tau|}\right)\right)\hat{F}_1(\tau)d\tau,$$

**5.20 Corollary** (Reconstruction of the boundary value of an $\vec{\alpha}$-holomorphic function by one of its bicomplex components) *Given one of the bicomplex components* $F_1$ *or* $F_2$ *of a boundary value* $f$ *of a function* $\tilde{f} \in \mathfrak{A}_{\vec{\alpha},p}(\mathbf{R}_3^+)$, $\vec{\alpha} = \alpha_1 i_1 + \alpha_2 i_2$, *the function* $f$ *is determined uniquely by*

$$f = H_{\vec{\alpha}}[F_2] + F_2 i_1 = F_1 + H_{\vec{\alpha}}^{-1}[F_1]i_1. \tag{5.46}$$

PROOF. This follows directly from (5.45). □

**5.21 Corollary** (Reconstruction of an $\vec{\alpha}$-holomorphic function by one of the bicomplex coordinates of its boundary value) *Given one of the bicomplex components* $F_1$ *or* $F_2$ *of the boundary value* $f$ *of a function* $\tilde{f} \in \mathfrak{A}_{\vec{\alpha},p}(\mathbb{R}_3^+)$, $\vec{\alpha} = \alpha_1 i_1 + \alpha_2 i_2$, *the function* $\tilde{f}$ *is determined uniquely by*

$$\tilde{f} = K_{\vec{\alpha}}[H_{\vec{\alpha}}[F_2] + F_2 i_1] = K_{\vec{\alpha}}[F_1 + H_{\vec{\alpha}}^{-1}[F_1] i_1].$$

**5.22** Now we proceed to the second most important class of $\alpha$-holomorphic functions with parameter $\alpha = \alpha_0 =: \gamma \in \mathbb{C}$. For definiteness we assume $\mathrm{Im}\gamma < 0$.

Consider $\gamma$-holomorphic functions, $\gamma \in \mathbb{C}$, $\mathrm{Im}\gamma < 0$. Let us fix for a given $\gamma$ the purely vectorial complex quaternion $\vec{\alpha}$ which satisfies $\gamma^2 = \vec{\alpha}^2$ and $\alpha_3 = 0$. Thus $\gamma^2 = -(\alpha_1^2 + \alpha_2^2)$.

As before, $\Gamma = \mathbb{R}^2$, $\vec{n} = -i_3$, $\mathbb{R}_3^+$ is the upper half-space, $P^{\pm} := \frac{1}{2\gamma} M^{\gamma \pm \vec{\alpha}}$.

By Theorem 5.9 and Remark 5.10, any function $u \in L_p(\Gamma; \mathbb{H}(\mathbb{C}))$ is $\gamma$-holomorphically extendable onto $\mathbb{R}_3^+$ iff $u \in \mathrm{im} P_\gamma$. Further, any $u$ is representable as

$$u = u^+ + u^- := P^+[u] + P^-[u].$$

Let us show that

$$u \in \mathrm{im} P_\gamma \iff u^{\pm} \in \mathrm{im} P_{\pm\vec{\alpha}}.$$

Indeed,

$$u \in \mathrm{im} P_\gamma \iff u = P_\gamma[u] \iff u = P^+ P_{\vec{\alpha}}[u] + P^- P_{-\vec{\alpha}}[u],$$

which is equivalent to the system

$$P^+[u] = P^+ P_{\vec{\alpha}}[u], \quad P^-[u] = P^- P_{-\vec{\alpha}}[u]. \tag{5.47}$$

The operators $P_{\pm\vec{\alpha}}$ and $P^{\pm}$ commute because they contain the same parameter $\vec{\alpha}$. Hence (5.47) is equivalent to

$$u^{\pm} = P_{\pm\vec{\alpha}}[u^{\pm}] \tag{5.48}$$

from which we see that $u^{\pm} \in \operatorname{im} P_{\pm\vec{\alpha}}$.

**5.23** Denote, as earlier, $U_1 := u_0 + u_3 i_3$, $U_2 := u_1 + u_2 i_3$ the bicomplex components of $u$, and let $U_2^{\pm} := u_1^{\pm} + u_2^{\pm} i_3 = (P^{\pm}[u])_1 + (P^{\pm}[u])_2 i_3$. Recall that for a quaternion $w$ we denote by $w_k = (w)_k$ its $k$th component in the standard representation $w = \sum_{k=0}^{3} w_k i_k$, and this should not be confused with the bicomplex components $W_k$. From Corollary 5.20 we conclude that

$$u = u^+ + u^- = H_{\vec{\alpha}}[U_2^+] + U_2^+ i_1 + H_{-\vec{\alpha}}[U_2^-] + U_2^- i_1 =$$

$$= H_{\vec{\alpha}}[U_2^+] + H_{-\vec{\alpha}}[U_2^-] + U_2 i_1$$

and thus

$$u \in \operatorname{im} P_{\gamma} \iff U_1 = H_{\vec{\alpha}}[U_2^+] + H_{-\vec{\alpha}}[U_2^-]. \tag{5.49}$$

Transforming the right side of (5.49) we get:

$$H_{\vec{\alpha}}[U_2^+] + H_{-\vec{\alpha}}[U_2^-] =$$

$$= \alpha_2 \check{V}_{\gamma}[U_2^+] - R_{\gamma,2}[\hat{U}_2^+] + \alpha_1 \check{V}_{\gamma}[U_2^+] i_3 + R_{\gamma,1}[\hat{U}_2^+] i_3 -$$

$$-\alpha_2 \check{V}_{\gamma}[U_2^-] - R_{\gamma,2}[\hat{U}_2^-] - \alpha_1 \check{V}_{\gamma}[U_2^-] i_3 + R_{\gamma,1}[\hat{U}_2^-] i_3 =$$

$$= -R_{\gamma,2}[\hat{U}_2^+ + \hat{U}_2^-] + \alpha_2 \check{V}_{\gamma}[U_2^+ - U_2^-] +$$

$$+ R_{\gamma,1}[\hat{U}_2^+ + \hat{U}_2^-] i_3 + \alpha_1 \check{V}_{\gamma}[U_2^+ - U_2^-] i_3. \tag{5.50}$$

But

$$\begin{aligned}
\hat{U}_2^+ + \hat{U}_2^- &= (\widehat{U_2^+ + U_2^-}) = \hat{U}_2, \\
U_2^+ - U_2^- &= ((P^+ - P^-)[u])_2 = \left(u \cdot \frac{\vec{\alpha}}{\gamma}\right)_2 = \\
&= \frac{1}{\gamma}(-U_2(\alpha_1 - \alpha_2 i_3) + U_1(\alpha_1 + \alpha_2 i_3)i_1)_2 = \\
&= \frac{1}{\gamma}U_1(\alpha_1 + \alpha_2 i_3).
\end{aligned}$$

Hence, (5.50) implies:

$$H_{\vec{\alpha}}[U_2^+] + H_{-\vec{\alpha}}[U_2^-] = -R_{\gamma,2}[U_2] + R_{\gamma,1}[\hat{U}_2]i_3 +$$

$$+\frac{\alpha_2}{\gamma}\check{V}_\gamma[U_1](\alpha_1 + \alpha_2 i_3) + \frac{\alpha_1}{\gamma}\check{V}_\gamma[U_1] \cdot (-\alpha_2 + \alpha_1 i_3). \tag{5.51}$$

Combining now (5.49) and (5.51) we arrive at the following assertion:

$$u \in \mathrm{im}P_\gamma \iff U_1 + \gamma \check{V}_\gamma[U_1]i_3 = R_{\gamma,1}[\hat{U}_2]i_3 - R_{\gamma,2}[\hat{U}_2]. \tag{5.52}$$

In a certain sense, (5.52) may be considered as the Hilbert formulas for $\gamma \in \mathbb{C}$ because we have $U_1$ only on the left-hand side, and $U_2$ only on the right-hand side. It is worth examining the equation (5.52) more carefully to see whether at least one of the operators defining both sides of (5.52) is invertible. We do this now.

It is remarkable that (5.52) does not already contain $\alpha$. This parameter played an auxiliary role only, and the result does not depend on $\alpha$.

**5.24** The equation (5.52) is equivalent to the system of two scalar integral equations

$$u_0(x) - 2\gamma \int_{\mathbb{R}^2} \theta_\gamma(x-y)u_3(y)dy = R_{\gamma,1}[u_2](x) - R_{\gamma,2}[u_1](x),$$

$$u_3(x) + 2\gamma \int_{\mathbf{R}^2} \theta_\gamma(x-y)u_0(y)dy = R_{\gamma,1}[u_1](x) + R_{\gamma,2}[u_2](x).$$

Rewrite them in the form

$$\begin{pmatrix} I & -2\gamma\theta_\gamma * \\ 2\gamma\theta_\gamma * & I \end{pmatrix} \begin{pmatrix} v_0 \\ v_1 \end{pmatrix} = \begin{pmatrix} c \\ d \end{pmatrix},$$

where $\theta_\gamma *$ denotes the convolution operator with the kernel $\theta_\gamma$;

$$v_0 := u_0; \quad v_1 := u_3;$$

$$\begin{aligned} c &:= R_{\gamma,1}[u_2] - R_{\gamma,2}[u_1]; \\ d &:= R_{\gamma,1}[u_1] - R_{\gamma,2}[u_2]. \end{aligned}$$

Let us multiply the last system from the left-hand side by the matrix

$$\Lambda := \begin{pmatrix} 1 & i \\ 1 & -i \end{pmatrix}.$$

Then we obtain

$$\begin{pmatrix} I + 2i\gamma\theta_\gamma * & 0 \\ 0 & I - 2i\gamma\theta_\gamma * \end{pmatrix} \begin{pmatrix} s_0 \\ s_1 \end{pmatrix} = \begin{pmatrix} \tilde{c} \\ \tilde{d} \end{pmatrix}, \tag{5.53}$$

where

$$\begin{pmatrix} s_0 \\ s_1 \end{pmatrix} := \Lambda \begin{pmatrix} v_0 \\ v_1 \end{pmatrix}; \quad \begin{pmatrix} \tilde{c} \\ \tilde{d} \end{pmatrix} := \Lambda \begin{pmatrix} c \\ d \end{pmatrix}.$$

Hence (5.52) is equivalent to a system of two convolution-type equations, each of them being of the form

$$w + \omega i \check{V}_\gamma[w] = \tilde{c}, \tag{5.54}$$

with the unknown $\mathbb{C}(i)$-valued function $w$, the given $\mathbb{C}(i)$-valued function $\tilde{c}$ and given constant $\omega \in \mathbb{C}(i)$. The left-hand side of (5.54) defines an operator

$$A_\omega := I + \omega \cdot i\check{V}_\gamma. \tag{5.55}$$

The peculiarities of the problems hint at the necessity to consider now some special functional spaces.

Let $\mathcal{S}$ denote, as usual, the Schwartz space of all complex-valued $\phi \in C^\infty(\mathbf{R}^2)$ such that

$$\sup_{x\in\mathbf{R}^2} |x^\mu \partial^\nu \phi(x)| < \infty$$

for any $\mu = (\mu_1, \mu_2)$ and $\nu = (\nu_1, \nu_2)$ where $\partial^\nu := \dfrac{\partial^{\nu_1+\nu_2}}{\partial x_1^{\nu_1} \partial x_2^{\nu_2}}$. For any $\phi \in \mathcal{S}$ its Fourier transform $\mathcal{F}[\phi]$ is defined by

$$\mathcal{F}[\phi](x) := \int_{\mathbf{R}^2} \phi(y) e^{ixy} dy.$$

Then

$$\mathcal{F}^{-1}[\phi](x) := \frac{1}{4\pi^2} \int_{\mathbf{R}^2} \phi(y) e^{-ixy} dy$$

is the inverse Fourier transform of $\phi$. Following [103] we introduce the Lizorkin space $\mathcal{L}$ of test functions. If

$$\mathcal{L}_0 := \{\psi \in \mathcal{S} | \partial^j \psi(0) = 0, \quad |j| = 0, 1, 2, \ldots\},$$

then $\mathcal{L}$ is the class dual to $\mathcal{L}_0$ consisting of the Fourier transforms of functions from $\mathcal{L}_0$:

$$\mathcal{L} := \mathcal{F}(\mathcal{L}_0) = \{\phi \in \mathcal{S} | \phi = \mathcal{F}[\psi], \psi \in \mathcal{L}_0\}.$$

This class admits a simple description: $\mathcal{L}$ consists precisely of those Schwartz functions $\psi$ which are orthogonal to all polynomials:

$$\int_{\mathbb{R}^2} x^j \psi(x)dx = 0, \quad |j| = 0, 1, 2, \ldots.$$

Let $S'$ be the space of the generalized Sobolev–Schwartz functions (= distributions). It is known (see, e.g., [121, p. 28]) that if $G \in S'$, $g \in \mathcal{L}$ then their convolution $G * g$ exists:

$$G * g(x) := \int_{\mathbb{R}^2} G(x-y))g(y)dy,$$

and

$$\mathcal{F}[G * g] = \mathcal{F}[G] \cdot \mathcal{F}[g].$$

Now consider the operator $A_\omega$ on $S$. Then

$$A_\omega = \mathcal{F}^{-1} \cdot B_\omega \cdot \mathcal{F} \tag{5.56}$$

where

$$B_\omega := 1 + 2\omega i \mathcal{F}[\theta_\gamma]$$

and where we identify a function and the corresponding multiplication operator. Thus the invertibility of the operator $A_\omega$ is equivalent to the invertibility of $B_\omega$.
First let us calculate the Fourier transform of the distribution $\theta_\gamma$.
The function $\theta_\gamma$ is spherically symmetric, so (see [86, p. 151], or [103, p. 352])

$$\mathcal{F}[\theta_\gamma](x) = \lim_{\epsilon \to 0}(-\frac{1}{2}\int_0^\infty e^{-(\epsilon+i\gamma)\rho} J_0(\rho|x|)d\rho),$$

where $J_0$ is the Bessel function of the first kind. We have assumed that $\mathrm{Im}\gamma < 0$ therefore $\mathrm{Re}(\epsilon + i\gamma) > 0$. Now, using the formula 794 from [21, p. 263], we obtain:

$$\mathcal{F}[\theta_\gamma] = -\frac{1}{2}(|x|^2 - \gamma^2)^{-1/2}. \tag{5.57}$$

Thus, finally,

$$B_\omega = 1 + 2\omega i(-\frac{1}{2})(|x|^2 - \gamma^2)^{-\frac{1}{2}} = \frac{(|x|^2-\gamma^2)^{1/2} - \omega i}{(|x|^2-\gamma^2)^{1/2}}. \tag{5.58}$$

We are considering the operator $A_\omega$ as acting on $\mathcal{S}$, hence $B_\omega$ acts on $\mathcal{S}$ also but is not in general invertible (of course, sometimes it is invertible; trivial example: $\omega = 0$).

Let us consider $B_\omega$ as acting on $\mathcal{L}_0$, so we have to consider $A_\omega$ on $\mathcal{L}$. This allows us to conclude the invertibility of $B_\omega$ on $\mathcal{L}_0$, which implies the invertibility of $A_\omega$ on $\mathcal{L}$. For our purposes, it is enough to consider the model cases $\omega = \pm\gamma$ (when $\omega = \gamma$ the numerator in (5.58) vanishes at $x = 0$).

Denoting $a := \frac{1}{i\gamma}$ we have:

$$B_{\pm\gamma}^{-1} = \frac{(|x|^2-\gamma^2)^{1/2}}{(|x|^2-\gamma^2)^{1/2} \mp \gamma i} =$$

$$= \frac{(|x|^2 + \frac{1}{a^2})^{1/2}}{(|x|^2+\frac{1}{a^2})^{1/2} \mp \frac{1}{a}} = \frac{(a^2|x|^2+1)^{1/2}}{(a^2|x|^2+1)^{1/2} \mp 1} =$$

$$= \frac{(a^2|x|^2+1)^{1/2}((a^2|x|^2+1)^{1/2} \pm 1)}{a^2|x|^2},$$

and thus

$$B_\gamma^{-1} = 1 + \frac{2}{a^2|x|^2} + \frac{(a^2|x|^2+1)^{1/2}-1}{a^2|x|^2}, \tag{5.59}$$

$$B_{-\gamma}^{-1} = 1 - \frac{(a^2|x|^2+1)^{1/2}-1}{a^2|x|^2}. \tag{5.60}$$

(5.56) gives now:

$$A_\gamma^{-1} = \mathcal{F}^{-1}(1 + \frac{2}{a^2|x|^2} + \frac{(a^2|x|^2+1)^{1/2}-1}{a^2|x|^2})\mathcal{F} =$$

$$= I + \mathcal{F}^{-1}[\frac{2}{a^2|x|^2}] * + \mathcal{F}^{-1}[\frac{(a^2|x|^2+1)^{1/2}-1}{a^2|x|^2}]*, \tag{5.61}$$

$$A_{-\gamma}^{-1} = I - \mathcal{F}^{-1}[\frac{(a^2|x|^2+1)^{1/2}-1}{a^2|x|^2}] * . \tag{5.62}$$

The definition of the mutually inverse Fourier transforms implies that

$$\mathcal{F}^{-1}[\frac{1}{|x|^2}] = (2\pi)^{-2}\mathcal{F}[\frac{1}{|x|^2}].$$

Considering the function $|x|^{-2}$ as a distribution over $\mathcal{L}$ and using a formula from [103, p. 361], we get

$$\mathcal{F}[\frac{1}{|x|^2}] = 2\pi ln\frac{1}{|x|}.$$

The convolution

$$2\pi\int_{\mathbf{R}^2} ln\frac{1}{|x-y|}\phi(y)dy =: R[\phi](x)$$

is the Riesz potential with respect to which the Lizorkin space is invariant; moreover, by [103, p. 364], $R(\mathcal{L}) = \mathcal{L}$. To treat the remaining terms in (5.61)-(5.62) we use the Bochner formula, see [103, p. 358]:

$$\mathcal{F}^{-1}[\frac{(a^2|x|^2+1)^{1/2}-1}{a^2|x|^2}] = \frac{1}{2\pi a^2}\int_0^\infty \frac{(a^2\rho^2+1)^{1/2}-1}{\rho}J_0(\rho|x|)d\rho =$$

$$= \frac{1}{2\pi a}(\int_0^\infty \frac{\rho}{(\rho^2+a^{-2})^{1/2}}J_0(\rho|x|)d\rho - \frac{1}{a}\int_0^\infty J_0(\rho|x|)d\rho +$$

$$+\frac{1}{a^2}\int_0^\infty (\rho^2+a^{-2})^{-1/2}J_0(\rho|x|)d\rho) =: \frac{1}{2\pi a}(I_1+I_2+I_3).$$

Now we apply three formulas from [1]: formula 11.4.44 gives

$$I_1 = \frac{1}{2|x|}exp(-\frac{|x|}{a}),$$

formula 11.4.33 gives

$$I_2 = -\frac{1}{a|x|},$$

and by formula 11.4.48 we have

$$I_3 = I_0(\frac{|x|}{2a}) \cdot K_0(\frac{|x|}{2a}),$$

where $I_0$ and $K_0$ are the Bessel functions of the second and third kinds respectively.

We combine all results obtained in this subsection in the following

**5.25 Proposition** *Let* $\gamma \in \mathbb{C}$, $\mathrm{Im}\gamma < 0$. *Then the operators* $A_{\pm\gamma} := I \pm \gamma i \check{V}_\gamma$ *acting on the Lizorkin space* $\mathcal{L}(\mathbb{R}^2, \mathbb{C})$ *are invertible and their inverses are determined by the formulas*

$$\begin{aligned} A_\gamma^{-1} &= I - \frac{\gamma^2}{2\pi^2} R + \frac{i\gamma}{2\pi}\left(\frac{e^{-i\gamma|x|}}{2|x|} - \frac{i\gamma}{|x|} + I_0(\frac{i\gamma|x|}{2\pi}) \cdot K_0(\frac{i\gamma|x|}{2})\right) *, \\ A_{-\gamma}^{-1} &= I - \frac{i\gamma}{2\pi}\left(\frac{e^{-i\gamma|x|}}{2|x|} - \frac{i\gamma}{|x|} + I_0(\frac{i\gamma|x|}{2}) \cdot K_0(\frac{i\gamma|x|}{2})\right) *, \end{aligned} \tag{5.63}$$

*where* $R$ *is the Riesz potential:* $R := 2\pi \ln\frac{1}{|x|} *$, $I_0$, $K_0$ *are the Bessel functions of the second and third kinds respectively; and for an appropriate function* $\phi$ *we write* $\phi *$ *to denote the convolution operator with the kernel* $\phi$.

Thus, for the operator $\mathbf{A}$ of the system (5.53),

$$\mathbf{A} := \begin{pmatrix} A_\gamma & 0 \\ 0 & A_{-\gamma} \end{pmatrix},$$

we have the inverse

$$\mathbf{A}^{-1} := \begin{pmatrix} A_\gamma^{-1} & 0 \\ 0 & A_{-\gamma}^{-1} \end{pmatrix}.$$

**5.26** In Subsection 5.23 we expressed the first bicomplex component of a function $u \in \operatorname{im} P_\gamma$ via its second one (formula (5.49)). By analogy, for the second bicomplex component we have

$$U_2 = H_{\vec{a}}^{-1}[U_1^+] + H_{-\vec{a}}^{-1}[U_1^-] \iff u \in \operatorname{im} P_\gamma. \tag{5.64}$$

Let us consider

$$H_{\vec{a}}^{-1}[U_1^+] + H_{-\vec{a}}^{-1}[U_1^-] =$$

$$= \alpha_2 \check{V}_\gamma[U_1^+] + R_{\gamma,2}[\hat{U}_1^+] - \alpha_1 \check{V}_\gamma[U_1^+]i_3 - R_{\gamma,1}[\hat{U}_1^+]i_3 -$$

$$-\alpha_2 \check{V}_\gamma[U_1^-] + R_{\gamma,2}[\hat{U}_1^-] + \alpha_1 \check{V}_\gamma[U_1^-]i_3 - R_{\gamma,1}[U_1^-]i_3 =$$

$$= R_{\gamma,2}[\hat{U}_1] + \alpha_2 \check{V}_\gamma[U_1^+ - U_1^-] - R_{\gamma,1}[\hat{U}_1]i_3 - \alpha_1 \check{V}_\gamma[U_1^+ - U_1^-]i_3. \tag{5.65}$$

For the difference $U_1^+ - U_1^-$ we have:

$$U_1^+ - U_1^- = -\frac{1}{\gamma} U_2(\alpha_1 - \alpha_2 i_3).$$

Hence, (5.65) implies that

$$\begin{aligned} H_{\vec{a}}^{-1}[U_1^+] + H_{-\vec{a}}^{-1}[U_1^-] &= R_{\gamma,2}[\hat{U}_1] - R_{\gamma,1}[\hat{U}_1]i_3 - \\ -\frac{\alpha_2}{\gamma} \check{V}_\gamma[U_2](\alpha_1 - \alpha_2 i_3) &+ \frac{\alpha_1}{\gamma} \check{V}_\gamma[U_2](\alpha_2 + \alpha_1 i_3). \end{aligned} \tag{5.66}$$

Combining (5.64) and (5.66) we arrive at the following assertion:

$$u \in \operatorname{im} P_\gamma \iff U_2 - \gamma \check{V}_\gamma[U_2]i_3 = -R_{\gamma,1}[\hat{U}_1]i_3 + R_{\gamma,2}[\hat{U}_1]. \tag{5.67}$$

The equation in (5.67) is equivalent to the system

$$u_1(x) + 2\gamma \int_{\mathbb{R}^2} \theta_\gamma(x-y) u_2(y) dy = -R_{\gamma,1}[u_2](x) + R_{\gamma,2}[u_1](x), \tag{5.68}$$

$$u_2(x) - 2\gamma \int_{\mathbb{R}^2} \theta_\gamma(x-y) u_1(y) dy = -R_{\gamma,1}[u_1](x) - R_{\gamma,2}[u_2](x), \tag{5.69}$$

or in matrix form

$$\begin{pmatrix} I & 2\gamma\theta_\gamma * \\ -2\gamma\theta_\gamma * & I \end{pmatrix} \begin{pmatrix} u_1 \\ u_2 \end{pmatrix} = \begin{pmatrix} f \\ g \end{pmatrix}, \tag{5.70}$$

where

$$f := -R_{\gamma,1}[U_2] + R_{\gamma,2}[U_1];$$

$$g := -R_{\gamma,1}[U_1] - R_{\gamma,2}[U_2].$$

Then, multiplying the system (5.70) on the left-hand side by the matrix

$$\tilde{\Lambda} := \begin{pmatrix} i & 1 \\ -i & 1 \end{pmatrix}$$

we obtain the system (5.53), where

$$\begin{pmatrix} s_0 \\ s_1 \end{pmatrix} := \tilde{\Lambda} \begin{pmatrix} u_1 \\ u_2 \end{pmatrix}; \begin{pmatrix} \tilde{c} \\ \tilde{d} \end{pmatrix} := \tilde{\Lambda} \begin{pmatrix} f \\ g \end{pmatrix}.$$

Then we may write down the analogues of the Hilbert formulas corresponding to the case $\alpha = \gamma \in \mathbb{C}$. Let us denote by $\Upsilon$ the mapping which assigns to a bicomplex function a $\mathbb{C}^2$-vector as follows:

$$\Upsilon[u_1 + u_2 i_3] =: \begin{pmatrix} u_1 \\ u_2 \end{pmatrix}.$$

**5.27 Definition** Let $\Gamma = \mathbb{R}^2$, $\vec{n} = -i_3$, $\gamma \in \mathbb{C}$. The operators

$$H_\gamma := \Upsilon^{-1}\Lambda^{-1}\mathrm{A}^{-1}\Lambda\Upsilon(M^{i_3}R_{\gamma,1} - R_{\gamma,2})\hat{Z}$$

and

$$H_\gamma^{-1} := \Upsilon^{-1}\tilde{\Lambda}^{-1}\mathrm{A}^{-1}\tilde{\Lambda}\Upsilon(-M^{i_3}R_{\gamma,1} + R_{\gamma,2})\hat{Z}$$

will be called "the $\gamma$-holomorphic Hilbert operators".

The reasons for this definition are justified by the next theorem.

Note that the mutual invertibility of the operators $H_\gamma^{\pm 1}$ follows from the equivalence of both (5.52) and (5.67) to the unique condition $u \in \operatorname{im} P_\gamma$.

The matrices $\Lambda^{-1}$, $\tilde{\Lambda}^{-1}$ occurring in the definition are given by

$$\Lambda^{-1} = \frac{1}{2}\begin{pmatrix} 1 & 1 \\ -i & i \end{pmatrix}; \quad \tilde{\Lambda}^{-1} = \frac{1}{2}\begin{pmatrix} -i & i \\ 1 & 1 \end{pmatrix}.$$

We will also use a more explicit form of the operators $H_\gamma$, $H_\gamma^{-1}$

$$H_\gamma = \frac{1}{2}(A_\gamma^{-1} + A_{-\gamma}^{-1} + i i_3(A_{-\gamma}^{-1} - A_\gamma^{-1}))(i_3 R_{\gamma,1} - R_{\gamma,2})\hat{Z},$$

$$H_\gamma^{-1} = -\frac{1}{2}(A_\gamma^{-1} + A_{-\gamma}^{-1} - i i_3(A_{-\gamma}^{-1} - A_\gamma^{-1}))(i_3 R_{\gamma,1} - R_{\gamma,2})\hat{Z}.$$

**5.28 Theorem** (Reconstruction of the boundary value of a $\gamma$-holomorphic function from one of its bicomplex components) *Given one of the bicomplex components $\{U_1, U_2\} \subset \mathcal{L}$ of a boundary value $u$ of a function $\tilde{u}$ $\gamma$-holomorphic in $\mathbb{R}^3_+$, $\gamma \in \mathbb{C}$, then the function $u$ is determined uniquely:*

$$u = H_\gamma[U_2] + U_2 i_1 = U_1 + H_\gamma^{-1}[U_1] i_1. \tag{5.71}$$

Note that

$$A_\gamma^{-1} + A_{-\gamma}^{-1} = 2I - \frac{\gamma^2}{2\pi^2}R,$$

and

$$A_{-\gamma}^{-1} - A_{\gamma}^{-1} = \frac{\gamma^2}{2\pi^2}R - \frac{i\gamma}{\pi}\left(\frac{e^{-i\gamma|x|}}{2|x|} - \frac{i\gamma}{|x|} + I_0(\frac{i\gamma|x|}{2\pi}) \cdot K_0(\frac{i\gamma|x|}{2})\right) * .$$

Hence formula (5.71) reads more explicitly as

$$u = \frac{1}{2}(2I - \frac{\gamma^2}{2\pi^2}R + ii_3(\frac{\gamma^2}{2\pi^2}R - \frac{i\gamma}{\pi}\left(\frac{e^{-i\gamma|x|}}{2|x|} - \frac{i\gamma}{|x|} + I_0(\frac{i\gamma|x|}{2\pi}) \cdot K_0(\frac{i\gamma|x|}{2})\right) *))$$

$$(i_3 R_{\gamma,1} - R_{\gamma,2})\hat{U}_2 + U_2 i_1.$$

**5.29 Corollary** (Reconstruction of a $\gamma$-holomorphic function from one of the bicomplex components of its boundary value) *Given one of the bicomplex components $\{U_1, U_2\} \subset \mathcal{L}$ of a boundary value $u$ of a function $\tilde{u}$ $\gamma$-holomorphic in $\mathbb{R}^3_+$ function $\tilde{u}$, then the function $u$ is determined uniquely:*

$$\tilde{u} = K_\gamma[H_\gamma[U_2] + U_2 i_1] = K_\gamma[U_1 + H_\gamma^{-1}[U_1] i_1].$$

5.30 REMARK In this section, especially in Subsections 5.12–5.21, we followed tradition and considered the main objects, in particular, the singular integral operator $S_{\vec{\alpha}}$, on the space $L_p(\mathbb{R}^2)$. But in many cases, including some physical applications, we need to work with less discontinuous functions, as a rule, with those satisfying the Hölder condition. The infinite domain $\mathbb{R}^2$ introduces some peculiarities into the situation. Nevertheless they can be taken care of, and the results of the above-mentioned subsections remain true if we consider either functions from $L_p(\mathbb{R}^2)$ satisfying the Hölder condition in the finite part of $\mathbb{R}^2$ or functions satisfying the Hölder condition in the finite part of $\mathbb{R}^2$ and vanishing sufficiently rapidly at infinity. Of course, we have to use then Lebesgue or Riemann integration respectively. On these functional classes the corresponding singular integrals exist in the sense of Cauchy's principal value (see, e.g., [135, Subsection 19.4], [34], compare with [38]).

Below, in Section 11 and Section 13 we shall use the assertions of Subsections 5.12–5.21 formulated for such $C^{0,\epsilon}$ without additional comments.

5.31 REMARK Both cases $\alpha = \vec{\alpha}$ and $\alpha = \alpha_0$ contain as a special instance the case $\alpha = 0$. The analogues of the Hilbert formulas in this particular case are well known (see, e.g., [15]). They can be easily derived from (5.45).

Moreover, for $\alpha = 0$ a special class of hyperholomorphic functions, the so-called Laplacian fields, are often considered in the literature and have various applications in mathematical physics. They are characterized by the inclusion $\varphi \in \ker D$ with the additional condition $\varphi_0 = 0$. It is obvious that the operator $S_0$ acts invariantly on purely vectorial functions $\vec{\varphi}$ if $\mathrm{Sc}(S_0[\vec{\varphi}]) = 0$. The latter condition is equivalent to

$$\int_\Gamma < \mathrm{grad}_x \frac{1}{|x-y|}, [\vec{n}(y) \times \vec{\varphi}(y)] > d\Gamma_y = 0. \tag{5.72}$$

Thus, on the space of purely vectorial Hölder functions satisfying additionally the equation (5.72), the operator of singular integration $S_0$ acts invariantly.

When $\vec{n} = -i_3$ the condition (5.72) reduces to $R_{0,1}[\varphi_2] - R_{0,2}[\varphi_1] = 0$. So, introducing the function space

$$X(\mathbf{R}^2) := \{\vec{\varphi} \in C^{0,\epsilon}(\Gamma) | R_{0,1}[\varphi_2] - R_{0,2}[\varphi_1] = 0\}$$

we obtain for functions $\vec{f}, \vec{g}$ from $X(\mathbf{R}^2)$ the pair of mutually inverse systems

$$\begin{cases} f_1 = -R_{0,1}[g_3], \\ f_2 = -R_{0,2}[g_3], \\ f_3 = -R_{0,1}[g_1] + R_{0,2}[g_2], \end{cases} \tag{5.73}$$

$$\begin{cases} g_1 = -R_{0,1}[f_3], \\ g_2 = -R_{0,2}[f_3], \\ g_3 = -R_{0,1}[f_1] + R_{0,2}[f_2], \end{cases} \tag{5.74}$$

which were given in [135, p. 144] and are called the three-dimensional integral Hilbert transforms. The English edition [136] of that book contains an appendix based on [137], which essentially refines the theory of the Hilbert transform for Laplacian vector fields (see also [107]).

# 6 Boundary value problems for $\alpha$-holomorphic functions

**6.1** The results of previous sections permit investigation of a number of boundary value problems for $\alpha$-holomorphic functions. First we consider the problem of $\alpha$-holomorphic extension of a $\mathbb{H}(\mathbb{C})$-valued function from the boundary onto the domain.

**6.2 Problem** (The Dirichlet problem for $(\psi, \alpha)$-holomorphic functions)

Given a structural set $\psi$ and complex quaternion $\alpha$, given an $\mathbb{H}(\mathbb{C})$-valued function $g$ on $\Gamma = \partial\Omega$, the problem is to find a function $f \in {}^{\psi}\mathfrak{A}_{\alpha}(\Omega)$ such that

$$f|_{\Gamma} = g. \tag{6.1}$$

**Solution**. As usual, the solution depends on $\Omega$, its boundary $\Gamma$, and the boundary function $g$ of $f$. We assume, naturally, that $\Gamma$ is a Liapunov surface or a hyperplane, and that $g \in C^{0,\epsilon}(\Gamma)$ or $g \in L_p(\Gamma)$.

By Theorem 5.7, Problem 6.2 is solvable if and only if

$$f = {}^{\psi}S_{\alpha}[f] \tag{6.2}$$

with ${}^{\psi}S_{\alpha}$ the corresponding singular integration operator with the quaternionic Cauchy kernel.

Moreover, if the condition of solvability (6.2) is fulfilled, then Problem 6.2 has the unique solution

$$f = {}^{\psi}K_{\alpha}[g]. \tag{6.3}$$

This solution covers the case of an unbounded domain $\Omega$ also, which includes automatically the condition of vanishing of $f$ at infinity. To conclude this subsection, we mention that the Dirichlet problem for holomorphic functions will be treated in Section 15.

**6.3** In Subsections 5.12–5.31, where we constructed the Hilbert operators and Hilbert formulas corresponding to $\alpha = \vec{\alpha}$ and to $\alpha = \alpha_0$, some important boundary value problems were treated. Here we are going to formulate them more explicitly.

Let $\mathbb{R}^2 = \Gamma = \{x \in \mathbb{R}^3 | x^3 = 0\}$ and $\vec{n} \equiv -i_3$. First let $\alpha = \vec{\alpha}$, that is, $\mathrm{Sc}(\alpha) = 0$. We assume additionally that $\alpha_3 = 0$. Thus, the complex quaternion $\alpha$ has the form $\alpha = \alpha_1 i_1 + \alpha_2 i_2$. As earlier, $\vec{\alpha}^2 = \gamma^2$, $\gamma \in \mathbb{C}$.

**6.4 Problem** (Finding an $\vec{\alpha}$-holomorphic function in a half-space from the boundary values of its two coordinates) Given two $\mathbb{C}$-valued functions $a, b$ on $\Gamma$, to find $f \in \mathfrak{A}_{\vec{\alpha}}(\mathbb{R}^3_+)$ such that

$$f_0|_\Gamma = a, \quad f_3|_\Gamma = b. \tag{6.4}$$

**Solution.** Again we assume that $\{a, b\} \subset C^{0,\epsilon}(\mathbb{R}^2)$ or $\{a, b\} \subset L_p(\mathbb{R}^2)$. Then Theorem 5.7 together with Theorem 5.19 shows that $f$ is defined uniquely, and the solution is constructed as follows.

First we construct the bicomplex function $G_1 := a + b i_3$ and the bicomplex function $G_2 := H^{-1}_{\vec{\alpha}}[G_1]$, where the operators $H^{\pm 1}_{\vec{\alpha}}$ are defined in Subsection 5.18. Then the solution to our problem is given by

$$f = K_{\vec{\alpha}}[g]$$

where $g := G_1 + G_2 i_1$.

If we change the boundary condition (6.4) to

$$f_1|_\Gamma = a, \quad f_2|_\Gamma = b \tag{6.5}$$

then $f = K_{\vec{\alpha}}[g]$ where $g = G_1 + G_2 i_1, G_2 = a + b i_3$, $G_1 = H_{\vec{\alpha}}[G_2]$.

**6.5** It is paradoxical, but the situation is more complicated in the case $\alpha = \alpha_0 =: \gamma \in \mathbb{C}$ (Subsections 5.22–5.30).

**6.6 Problem** (Finding a $\gamma$-holomorphic function in a half-space from the boundary value of its two components) Given two $\mathbb{C}$-valued functions on $\mathbb{R}^2$, to find $f \in \mathfrak{A}_\gamma(\mathbb{R}^3_+)$ such that

$$f_0|_\Gamma = a, \quad f_3|_\Gamma = b.$$

**Solution.** The considerations of Subsections 5.22 - 5.30 force us to assume that $\{a, b\} \subset \mathcal{L}(\Gamma)$, the Lizorkin space (see Subsection 5.24). Then $f = K_\gamma[g]$ where $g = G_1 + G_2 i_1$, $G_2 = a + b i_3$ and $G_1 = H_\gamma[G_2]$ with $H_\gamma$ defined in Subsection 5.27.

**6.7** The technique at our disposal now allows us to investigate many other problems for hyperholomorphic functions. We give one more example only, an analogue of the famous Riemann boundary value problem.

The simplest version of it reads as follows. Let $\Gamma$ be a closed Liapunov surface in $\mathbb{R}^3$ bounding a domain $\Omega^+$. Given $g \in C^{0,\epsilon}(\Gamma; \mathbb{H}(\mathbb{C}))$ and constant $G \in \mathbb{H}(\mathbb{C})$, to find a pair of functions $f^\pm \in \mathfrak{A}_\gamma(\Omega^\pm)$, $\Omega^- := \mathbb{R}^3 \setminus \overline{\Omega^+}$, with the boundary condition

$$f^+(y) = f^-(y)G + g(y), \quad y \in \Gamma. \tag{6.6}$$

For $\gamma \in \mathbb{C}$, the problem can be reduced immediately to the Plemelj–Sokhotski formulas by denoting $f^- \cdot G =: \tilde{f}^-$. But we shall demonstrate another useful way to solve it. Assume additionally that $G \notin \mathfrak{S} \cup \{0\}$, and let us introduce

$$\check{G}(x) := \begin{cases} G, & x \in \Omega^+, \\ 1, & x \in \Omega^-. \end{cases} \tag{6.7}$$

$\check{G}$ satisfies the boundary condition

$$\check{G}^+(y) = \check{G}^-(y)G, \quad y \in \Gamma, \tag{6.8}$$

i.e.,

$$\check{G}^+(y) = G, \quad y \in \Gamma. \tag{6.9}$$

(6.9) and (6.6) imply that

$$f^+ \cdot (\check{G}^+)^{-1} = f^- + g \cdot (\check{G}^+)^{-1}. \tag{6.10}$$

For the latter problem its unique solution is given by

$$f(x) = K_\gamma[g \cdot G^{-1}](x) \cdot \check{G}(x). \tag{6.11}$$

**6.8** In [111] the Fredholm theory for the different generalizations of the Riemann boundary value problems for $\alpha = 0$ was constructed. Non-commutativity of quaternionic multiplication resulted in two direct analogues for each structural set $\psi$.

The "right" Riemann boundary value problems: to find a pair of functions $f^\pm$ from the classes ${}^{\psi}\mathfrak{A}_0(\Omega^\pm; X)$ (see Subsection 5.8) satisfying the boundary condition on $\Gamma$:

$$f^+ a = f^- b + g, \tag{6.12}$$

where $g \in X$, and $a$ and $b$ are given multipliers for $X$. If we change the boundary condition (6.12) to

$$af^+ = bf^- + g \tag{6.13}$$

then we obtain the "left" Riemann b.v.p. problem.

Having applied the quaternionic Plemelj–Sokhotski formulas one can reduce problems (6.12), (6.13) in a classical way to singular integral equations defined by the operators:

$$M^a P_0 + M^b Q_0; \quad {}^aMP_0 + {}^bMQ_0. \tag{6.14}$$

For an arbitrary $\alpha \in \mathbb{H}(\mathbb{C})$ we can state the same boundary value problems with the requirement that $f^\pm$ be in classes ${}^{\psi}\mathfrak{A}_\alpha(\Omega^\pm; X)$. Then they reduce in the same manner to singular integral equations defined by the operators:

$$M^a \cdot {}^{\psi}P_\alpha + M^b \cdot {}^{\psi}Q_\alpha; \quad {}^aM \cdot {}^{\psi}P_\alpha + {}^bM \cdot {}^{\psi}Q_\alpha. \tag{6.15}$$

The Fredholm theory for the operators (6.14), as well as a description of the algebra generated by them and by multiplication operators by piecewise continuous functions, are given in detail in [111, 112]. Here we show only that the Fredholm theory for the operators (6.15) in essence coincides with that of the operators (6.14). For this purpose it is enough to prove that the operator ${}^{\psi}S_\alpha$ coincides with ${}^{\psi}S_0$ up to a compact operator.

Let us examine first the case $\alpha = \alpha_0 \in \mathbb{C}$. Then using the Taylor expansion for the function $e^{-i\alpha|x|}$ we have

$$({}^{\psi}\mathcal{S}_\alpha[f])(x) = -2\int_\Gamma \{-\frac{\alpha e^{-i\alpha|x-y|}}{4\pi|x-y|}\} + i\alpha\mathcal{K}_{\psi,0}(x-y)|x-y|e^{-i\alpha|x-y|}+$$

$$+\mathcal{K}_{\psi,0}(x-y)\sum_{k=1}^{\infty}\frac{(-i\alpha|x-y|)^k}{k!}\}\vec{n}_\psi(y)f(y)d\Gamma_y+$$

$$+({}^{\psi}\mathcal{S}_0[f])(x). \tag{6.16}$$

The kernel of the first integral in (6.16) is weakly singular; therefore this integral represents a compact operator (see, e.g. [81, p. 57]). Thus, for the case $\alpha \in \mathbb{C}$ we obtain

$${}^{\psi}\mathcal{S}_\alpha - {}^{\psi}S_0 = T, \tag{6.17}$$

where $T$ is a compact operator.

**6.9** The proof of (6.17) in cases

1) $\alpha \notin \mathfrak{S}$, $\vec{\alpha}^2 \neq 0$

and

3) $\alpha \in \mathfrak{S}$, $\alpha_0 \neq 0$

is obvious because for the operators $S_{\xi_+}, S_{\xi_-}, S_{2\alpha_0}$ which appear in the definition of $S_\alpha$ the equality (6.17) has been proved, and the projection operators $P^+$, $P^-$ commute with the operators of singular integration.

When $\alpha \in \mathfrak{S}$, $\alpha_0 = 0$ (case 4), the operator $\hat{V}_0$ is compact. Hence, in this case (6.17) is also true.

It remains to examine case 2) $\alpha \notin \mathfrak{S}$, $\vec{\alpha}^2 = 0$. A simple calculation gives

$$\frac{\partial}{\partial \alpha_0}({}^{\psi}S_{\alpha_0}[f])(x) = -2\int_\Gamma (1 - i\alpha_0|x-y| + \alpha_0(x-y))\theta_{\alpha_0}(x-y)\vec{n}_\psi(y)f(y)d\Gamma_y.$$

The kernel of this operator is weakly singular; therefore the operator is compact. Thus, for an arbitrary $\alpha \in \mathbb{H}(\mathbb{C})$ the equality (6.17) is true.

# 7 Vectorial reformulation of some complex–quaternionic objects and results

**7.1** Vector notation is most commonly used in electrodynamics, some models from which will be examined in the next chapter. Therefore to simplify the reading for specialists who are not accustomed to biquaternionic language, we reformulate here some objects and results from the previous sections in vector terms.

**7.2** In formulas (4.9)–(4.11) we introduced the important integral operators $T_{\alpha_0}$, $K_{\alpha_0}$, $S_{\alpha_0}$, $\alpha_0 \in \mathbb{C}$. Let us examine what they are in vector language. The kernel $\mathcal{K}_{\alpha_0}$ by definition is representable in the form

$$\mathcal{K}_{\alpha_0} = -D_{-\alpha_0}[\theta_{\alpha_0}],$$

where $\theta_{\alpha_0}(x) := -\dfrac{e^{-i\alpha_0|x|}}{4\pi|x|}$. Then we can rewrite the operator $T_{\alpha_0}$ as

$$T_{\alpha_0}[f](x) = \int_\Omega \{\alpha_0\theta_{\alpha_0}(x-y)f_0(y) + < \text{grad}_x\theta_{\alpha_0}(x-y), \vec{f}(y) >\}dy +$$

$$+\int_{\Omega}\{\alpha_0\theta_{\alpha_0}(x-y)\vec{f}(y)-\mathrm{grad}_x\theta_{\alpha_0}(x-y)\cdot f_0(y)\}dy. \tag{7.1}$$

The first integral in (7.1) represents the scalar part of the operator $T_{\alpha_0}$, the second integral gives the vector part.

**7.3** For the operator $K_{\alpha_0}$ we have a more complicated representation:

$$K_{\alpha_0}[f](x)=\int_{\Gamma}\{\alpha_0\theta_{\alpha_0}(x-y)<\vec{n}(y),\vec{f}(y)>-<\mathrm{grad}_x\theta_{\alpha_0}(x-y),\vec{n}(y)>f_0(y)-$$

$$-<[\mathrm{grad}_x\theta_{\alpha_0}(x-y)\times\vec{n}(y)],\vec{f}(y)>\}d\Gamma_y+\int_{\Gamma}\{-\alpha_0\theta_{\alpha_0}(x-y)[\vec{n}(y)\times\vec{f}(y)]-$$

$$-\alpha_0\theta_{\alpha_0}(x-y)\vec{n}(y)f_0(y)-<\mathrm{grad}_x\theta_{\alpha_0}(x-y),\vec{n}(y)>\vec{f}(y)+$$

$$+[\mathrm{grad}_x\theta_{\alpha_0}(x-y)\times\vec{n}(y)]f_0(y)+[[\mathrm{grad}_x\theta_{\alpha_0}(x-y)\times\vec{n}(y)]\times\vec{f}(y)]\}d\Gamma_y. \tag{7.2}$$

Again the first integral in (7.2) is the scalar part of $K_{\alpha_0}[f]$ and the second integral is the vector part.

**7.4** To understand better what the formulas (4.31), (4.32) are, let us consider the special case $\lambda\in\mathbb{C}$, $\alpha=\nu\in\mathbb{C}$. Let us also restrict our consideration to the purely vectorial $\mathbb{H}(\mathbb{C})$-valued function $f\equiv\vec{f}$. Then for $\vec{f}\in\ker\Delta_\lambda$ from (4.31) by the definitions of the operators $K_{\pm\alpha}$, $\Pi_{\pm\alpha}$ we obtain:

$$\vec{f}(x)=\frac{1}{2\nu}\int_{\Gamma}\{\mathrm{grad}_x\theta_\nu(x-y)\vec{n}(y)[(-D+\nu)\vec{f}(y)]-$$

$$-\nu\theta_\nu(x-y)\vec{n}(y)[(-D+\nu)\vec{f}(y)]+\mathrm{grad}_x\theta_\nu(x-y)\vec{n}(y)[(D+\nu)\vec{f}(y)]+$$

$$+\theta_\nu(x-y)\vec{n}(y)[(D+\nu)\vec{f}(y)]\}d\Gamma_y, x \in \Omega.$$

Consequently,

$$\vec{f}(x) = \int_\Gamma \{\text{grad}_x \theta_\nu(x-y)\vec{n}(y)\vec{f}(y) + \theta_\nu(x-y)\vec{n}(y)[D[\vec{f}](y)\}d\Gamma_y, \quad x \in \Omega. \tag{7.3}$$

Let us rewrite (7.3) in the vectorial form separating the scalar and the vectorial parts of the equality. We obtain

$$0 = \int_\Gamma \{< \text{grad}_x \theta_\nu(x-y), [\vec{n}(y) \times \vec{f}(y)] > +$$

$$\theta_\nu(x-y) < \vec{n}(y), \text{rot}\vec{f}(y) > \} d\Gamma_y, \quad x \in \Omega, \tag{7.4}$$

$$\vec{f}(x) = \int_\Gamma \{[\text{grad}_x \theta_\nu(x-y) \times [\vec{n}(y) \times \vec{f}(y)]] - \text{grad}_x \theta_\nu(x-y) < \vec{n}(y), \vec{f}(y) > -$$

$$-\theta_\nu(x-y)([\text{rot}\vec{f}(y) \times \vec{n}(y)] + \vec{n}(y)\text{div}\vec{f}(y))\}d\Gamma_y, \quad x \in \Omega. \tag{7.5}$$

The equality (7.5) represents the integral criterion for the function $\vec{f}$ to belong to the kernel of the Helmholtz operator (see [27, p. 118]). The equality (7.4) is the following form of the Gauss theorem from [91] (see also [27, p. 60]):

$$\int_\Gamma (\varphi \text{Div}\vec{a})d\Gamma + \int_\Gamma < \text{Grad}\varphi, \vec{a} > d\Gamma = 0, \tag{7.6}$$

where Div, Grad are surface operators (see the definition, e.g. in [27, p. 33, p. 60]). This can be seen if we take $\theta_\nu$ as $\varphi$ and $\vec{a} = [\vec{n} \times \vec{f}]$ and take into account that

$$\mathrm{Grad}_\nu \theta_\nu(x-y) = -\mathrm{grad}_x \theta_\nu(x-y) - \vec{n}(y)\frac{\partial \theta_\nu(x-y)}{\partial \vec{n}(y)},$$

$$\mathrm{Div}[\vec{n} \times \vec{f}] = - < \vec{n}, \mathrm{rot} \vec{f} >$$

([27, p. 61]). That is, (7.4) is satisfied identically.

# Chapter 2

# Electrodynamical models

## 8 The classical Maxwell equations

**8.1** Let $(\vec{E}, \vec{H})$ denote an electromagnetic field in a vacuum with $\vec{E} : \Xi \subset \mathbb{R}^4 \longrightarrow \mathbb{R}^3$ and $\vec{H} : \Xi \subset \mathbb{R}^4 \longrightarrow \mathbb{R}^3$ being respectively its electric and magnetic components. Here $\Xi = I \times \Omega$ with $I$ some interval in $\mathbb{R}$ and $\Omega$ some domain in $\mathbb{R}^3$. Then $\vec{E}$ and $\vec{H}$ satisfy the classical Maxwell equations:

$$\begin{aligned} \operatorname{div}\vec{E} = \operatorname{div}\vec{H} = 0, \\ \operatorname{rot}\vec{E} = -\partial_t\vec{H}, \quad \operatorname{rot}\vec{H} = \partial_t\vec{E}, \end{aligned} \tag{8.1}$$

where $\partial_t := \frac{\partial}{\partial t}, t \in I$ is the time variable, and the operations rot and div act with respect to the space variables $x_1, x_2, x_3$.

There exists a well-known (see, e.g., [116, 22, 51, 97, 52, 8]) reformulation of the system (8.1) in terms of complex quaternions. It arises naturally: if we introduce the purely vectorial function $\vec{f} := \vec{E} + i\vec{H}$ then it satisfies (compare with (2.5)–(2.6)) the following quaternionic equation

$$(\partial_t + iD)\vec{f} = 0, \tag{8.2}$$

where $D$ is the Moisil–Theodoresco operator with respect to the space variables.

**8.2** In Subsection 3.24 we introduced the notion of the Füter type operator $\partial_t - aD$ with $a = \text{Const} \in \mathbb{C}$. It appears that this operator describes some important properties of interesting physical objects: in Section 16 we shall examine its applications in the theory of instantons; in Section 12 we shall encounter the fact that the set $\ker(\partial_t - iD)$ is very closely and simply related to the Dirac equation describing a spinor field corresponding to a neutrino.

Here with respect to the Maxwell equations we consider in addition to (8.2) the equation

$$(\partial_t - iD)F = 0, \tag{8.3}$$

and we establish a connection between its solutions and electromagnetic potentials.

Let us introduce in the standard way the electromagnetic potentials $\phi, \vec{A}$:

$$\vec{E} = -\mathrm{grad}\phi - \partial_t\vec{A}, \quad \vec{H} = \mathrm{rot}\vec{A}, \tag{8.4}$$

where $\vec{E}$ and $\vec{H}$ are solutions of (8.1).

Having substituted (8.4) into (8.1) we get the equalities:

$$\mathrm{rot}\,\mathrm{rot}\vec{A} + \partial_t^2\vec{A} + \partial_t\mathrm{grad}\phi = 0, \tag{8.5}$$

$$\partial_t\mathrm{div}\vec{A} + \Delta\phi = 0, \tag{8.6}$$

which are characteristic for electromagnetic potentials. In order to diminish the remaining arbitrariness in the choice of $\phi$ and $\vec{A}$, it is usual to introduce some additional conditions, the so–called gauge conditions (see, e.g., [54, Subsection 1.10], [39]). The most commonly used is the Lorentz gauge

$$\mathrm{div}\vec{A} + \partial_t\phi = 0. \tag{8.7}$$

Minkowski proposed combining the electromagnetic potentials $\phi$ and $\vec{A}$ and considering the following vector of the four-dimensional world: $F := (i\phi, \vec{A})$ (see [96, p.114]).

We now show that each solution of (8.3) is a Minkowski vector in the Lorentz gauge.

**8.3 Proposition** *Let* $\phi : \Xi \longrightarrow \mathbb{C}$, $\vec{A} : \Xi \longrightarrow \mathbb{C}^3$, *and*

$$F := (i\phi, \vec{A}) = i\phi + \sum_{k=1}^{3} A_k i_k \in C^2(\Xi) \cap \ker(\partial_t - iD).$$

*Then $\phi$ and $\vec{A}$ are electromagnetic potentials in the Lorentz gauge; i.e., the equalities (8.5)–(8.7) are valid.*

PROOF. In vector form, the condition $F \in \ker(\partial_t - iD)$ can be rewritten as follows:

$$i\partial_t\phi + i\mathrm{div}\vec{A} = 0, \tag{8.8}$$

$$\partial_t\vec{A} + \mathrm{grad}\phi - i\mathrm{rot}\vec{A} = 0. \tag{8.9}$$

Thus, the scalar equality (8.8) gives (8.7) and it remains to show (8.5), (8.6). Applying the operator div to (8.9) we get (8.6) and applying the operator $\partial_t$ to (8.9) we obtain

$$\partial_t^2\vec{A} + \partial_t\mathrm{grad}\phi - i\partial_t\mathrm{rot}\vec{A} = 0. \tag{8.10}$$

From (8.9) it follows that

$$i\partial_t\mathrm{rot}\vec{A} = -\mathrm{rotrot}\vec{A}.$$

Then from (8.10) we have (8.5). □

**8.4** It is known that both the electromagnetic field components $\vec{E}, \vec{H}$ and the electromagnetic potentials $\phi, \vec{A}$ belong to the null set of the d'Alambertian $\partial_t^2 - \Delta_{\mathbb{R}^3}$. Properties of the complex quaternions allow us to factorize it by means of two "conjugate" Füter type operators:

$$\partial_t^2 - \Delta_{\mathbb{R}^3} = (\partial_t - iD)(\partial_t + iD). \tag{8.11}$$

Thus, it is not merely by chance that the operators $\partial_t + iD$ and $\partial_t - iD$ arise in (8.2) and in Proposition 8.3.

**8.5** Because of the linearity of the Maxwell equations (8.1), an electromagnetic field arbitrarily changing in time can be represented as a sum (finite or infinite) of time–harmonic (monochromatic) fields. The dependence on time for time–harmonic fields is expressed by the factor $e^{-i\omega t}$. The Maxwell equations describing the time–harmonic electromagnetic fields in a homogeneous, isotropic region devoid of currents and charges can be written down as follows. Each component of an electromagnetic field is characterized now by a frequency $\omega \in \mathbb{R}$ and by a complex amplitude, i.e., $\vec{E} = e^{-i\omega t}\vec{E}_c(x)$, $\vec{H} = e^{-i\omega t}\vec{H}_c(x)$ with $\vec{E}_c : \Omega \to \mathbb{C}^3$, $\vec{H}_c : \Omega \to \mathbb{C}^3$. From now on we omit the subindex "$c$" and we write $\vec{E}$ and $\vec{H}$ for the amplitudes. Then the corresponding Maxwell system has the form

$$\mathrm{rot}\vec{E} = i\omega\mu\vec{H}, \quad \mathrm{rot}\vec{H} = \sigma\vec{E} \quad \text{in } \Omega, \tag{8.12}$$

where $\{\mu, \sigma\} \subset \mathbb{C}$, $\mu$ and $\sigma$ are constants characterizing the medium ( $\mu$ is a complex magnetic permeability, $\sigma$ is a complex electrical conductivity). More details can be found, e.g., in [48, 54]. The domain $\Omega$ is assumed to have a Liapunov boundary $\Gamma := \partial\Omega$; $\vec{E}, \vec{H} : \Omega \to \mathbb{C}^3$.

**8.6** Let us rewrite the equations (8.12) in matrix form:

$$\begin{pmatrix} \mathrm{rot} & -i\omega\mu \\ -\sigma & \mathrm{rot} \end{pmatrix} \begin{pmatrix} \vec{E} \\ \vec{H} \end{pmatrix} = 0, \tag{8.13}$$

and define an operator

$$\mathcal{M} := \begin{pmatrix} \mathrm{rot} & -i\omega\mu \\ -\sigma & \mathrm{rot} \end{pmatrix}.$$

Its natural domain is $C^1(\Omega; \mathbb{C}^3 \times \mathbb{C}^3)$. Note that from (8.12), or from (8.13), it follows that every solution of (8.13) has the property

$$\mathrm{div}\vec{E} = \mathrm{div}\vec{H} = 0. \tag{8.14}$$

Taking into account (8.14) let us introduce for $k \in \mathbb{N}$

$$\hat{C}^k := \hat{C}^k(\Omega; \mathbb{C}^3 \times \mathbb{C}^3) := \{(\vec{f}, \vec{g}) \mid (\vec{f}, \vec{g}) \in C^k(\Omega; \mathbb{C}^3 \times \mathbb{C}^3), \mathrm{div}\vec{f} = \mathrm{div}\vec{g} = 0\}.$$

It is essential to note that $\mathcal{M}$ maps a solenoidal (=divergenceless) vector into a solenoidal one reducing, of course, the smoothness: if $(\vec{f}, \vec{g}) \in \hat{C}^{k+1}$ then $\mathcal{M}[(\vec{f}, \vec{g})] \in \hat{C}^k$.

**8.7** The necessary conditions (8.14) allow us to rewrite (8.13) in the form

$$\mathcal{M}_{\mathbb{H}}\begin{pmatrix}\vec{E}\\ \vec{H}\end{pmatrix}=0, \tag{8.15}$$

where

$$\mathcal{M}_{\mathbb{H}}:=\begin{pmatrix}D & -i\omega\mu\\ -\sigma & D\end{pmatrix}. \tag{8.16}$$

It is natural to consider $\mathcal{M}_{\mathbb{H}}$ not only on purely vectorial complex–quaternionic functions but to extend it to pairs of complete complex quaternions using the same notation:

$$\mathcal{M}_{\mathbb{H}}: C^1(\Omega;\mathbb{H}(\mathbb{C})\times\mathbb{H}(\mathbb{C}))\longrightarrow C(\Omega;\mathbb{H}(\mathbb{C})\times\mathbb{H}(\mathbb{C})).$$

We shall call this operator the quaternionic Maxwell operator. On purely vectorial complex–quaternionic functions the quaternionic Maxwell operator coincides with the usual Maxwell operator. We shall sometimes call functions from $\ker\mathcal{M}_{\mathbb{H}}$ the "quaternionic Maxwell functions".

# 9 Relationship between time–harmonic electromagnetic fields and $\alpha$-holomorphic functions

**9.1** Let $\lambda := i\omega\mu\sigma$ , and let $\alpha\in\mathbb{C}$ be a square root of $\lambda$: $\alpha^2=\lambda$ (the case $\alpha\in\mathbb{H}(\mathbb{C})$ with respect to the Maxwell equations will be considered below in Subsection 9.6 and Proposition 9.7). In this situation the functions $\vec{E}$ and $\vec{H}$, components of an electromagnetic field, satisfy the Helmholtz equation:

$$\begin{pmatrix}\Delta_\lambda & 0\\ 0 & \Delta_\lambda\end{pmatrix}\begin{pmatrix}\vec{E}\\ \vec{H}\end{pmatrix}=0. \tag{9.1}$$

This can be seen from (8.13) after applying the operator

$$\begin{pmatrix} -D & -i\omega\mu \\ -\sigma & -D \end{pmatrix}.$$

The set $\ker \Delta_\lambda$ contains not only solutions of Maxwell's equations. If

$$\mathbb{M}(\Omega) := \{\vec{u} \mid \vec{u} \in \ker \Delta_\lambda, \ \operatorname{div}\vec{u} = 0 \ \text{ in } \ \Omega\},$$

then we have obtained a class of functions which generates solutions to the Maxwell equations (8.12) in the following sense: for any $\vec{E} \in \mathbb{M}(\Omega)$ the pair

$$(\vec{E}; \quad \vec{H} := \frac{1}{i\omega\mu}\operatorname{rot}\vec{E})$$

is a solution to (8.12).

The following proposition shows how to construct the projection operator onto the set $\mathbb{M}(\Omega)$.

**9.2 Proposition** *The operators*

$$P := -\frac{1}{\alpha^2}\operatorname{graddiv}; \quad R := \frac{1}{\alpha^2}\operatorname{rotrot}$$

*are mutually complementary projectors on* $\ker(\Delta + \alpha^2 I)$. *Moreover,* $\vec{u} \in \mathbb{M}(\Omega)$ *iff* $\vec{u} \in \operatorname{im}R$ *(or, which is equivalent,* $\vec{u} \in \ker P$ *).*

PROOF. Firstly, let us show that $P$ and $R$ are mutually complementary projectors on $\ker(\Delta + \alpha^2 I)$. Let $\vec{u} \in \ker(\Delta + \alpha^2 I)$. Then

$$\vec{u} = -\frac{1}{\alpha^2}\Delta\vec{u}.$$

Consequently,

$$\vec{u} = \frac{1}{\alpha^2}\operatorname{rotrot}\vec{u} - \frac{1}{\alpha^2}\operatorname{graddiv}\vec{u}.$$

That is,

$$\vec{u} = P[\vec{u}] + R[\vec{u}].$$

Besides

$$PR[\vec{u}] = RP[\vec{u}] = 0.$$

Hence

$$P^2[\vec{u}] = P(P+R)[\vec{u}] = P[\vec{u}];$$

$$R^2[\vec{u}] = R(P+R)[\vec{u}] = R[\vec{u}].$$

Let us verify the equivalence of the conditions $\vec{u} \in \mathbb{M}(\Omega)$ and $\vec{u} \in \mathrm{im}R$. The second inclusion follows evidently from the first one by the definition of the operator $P$.

Conversely, if $\vec{u} \in \mathrm{im}R$, then $P[\vec{u}] = 0$. Hence, $\mathrm{div}\vec{u} = c$, where $c = \mathrm{Const}$. But if $\vec{u} \in \ker(\Delta + \alpha^2 I)$ then $\mathrm{div}\vec{u} \in \ker(\Delta + \alpha^2 I)$ also. The kernel of the Helmholtz operator $\Delta + \alpha^2 I$ does not contain constants. Therefore $\mathrm{div}\vec{u} = 0$ and $\vec{u} \in \mathbb{M}(\Omega)$.

□

From the vectorial representation (see Subsection 3.11) of the operator $D_\alpha$ we get another characterization of the set $\mathbb{M}(\Omega)$.

**9.3 Proposition** *Let* $\vec{u} \in \ker(\Delta + \alpha^2 I)$. *Then* $\vec{u} \in \mathbb{M}(\Omega)$ *if and only if* $\mathrm{Sc}(\Pi_{-\alpha}\vec{u}) = \mathrm{Sc}(\Pi_\alpha \vec{u}) = 0$.

**9.4** Let $\{\vec{E}, \vec{H}\} \subset \ker \mathcal{M}$, then an easy calculation gives the following:

$$2\sigma\Pi_{-\alpha}[\vec{E}] = \sigma\vec{E} + \alpha\vec{H}; \quad 2\sigma\Pi_{\alpha}[\vec{E}] = \sigma\vec{E} - \alpha\vec{H};$$

$$2\alpha\Pi_{-\alpha}[\vec{H}] = \sigma\vec{E} + \alpha\vec{H}; \quad 2\alpha\Pi_{\alpha}[\vec{H}] = -\sigma\vec{E} + \alpha\vec{H}. \tag{9.2}$$

Let us write also $\vec{\varphi} := \sigma\vec{E} + \alpha\vec{H}$, $\vec{\psi} := -\sigma\vec{E} + \alpha\vec{H}$. The operators $\Pi_\alpha, \Pi_{-\alpha}$ are projectors onto $\ker D_\alpha, \ker D_{-\alpha}$ correspondingly (see Proposition 2.9 and Theorem 2.13). Therefore from (9.2) it follows that

$$\varphi \in \ker D_{-\alpha}; \quad \psi \in \ker D_\alpha.$$

That is, for $\vec{E}, \vec{H}$ satisfying (8.12) we have:

$$\begin{pmatrix} D_{-\alpha} & 0 \\ 0 & D_{\alpha} \end{pmatrix} B_{\alpha} \begin{pmatrix} \vec{E} \\ \vec{H} \end{pmatrix} = 0, \tag{9.3}$$

where

$$B_{\alpha} := \begin{pmatrix} \sigma & \alpha \\ -\sigma & \alpha \end{pmatrix}.$$

The matrix $B_{\alpha}$ is invertible, with inverse

$$B_{\alpha}^{-1} = \begin{pmatrix} \frac{1}{2\sigma} & -\frac{1}{2\sigma} \\ \frac{1}{2\alpha} & \frac{1}{2\alpha} \end{pmatrix}.$$

Then for the quaternionic Maxwell operator we have the equality

$$\mathcal{M}_{\mathbb{H}} = B_{\alpha}^{-1} \begin{pmatrix} D_{-\alpha} & 0 \\ 0 & D_{\alpha} \end{pmatrix} B_{\alpha}; \tag{9.4}$$

that is, we have "diagonalized" the operator $\mathcal{M}_{\mathbb{H}}$. For the "usual" Maxwell operator $\mathcal{M}$ we should remember that it acts on pairs of purely vectorial complex–quaternionic functions. But for $(\vec{E}, \vec{H}) \in \hat{C}^1$ we have the equality

$$\mathcal{M} \begin{pmatrix} \vec{E} \\ \vec{H} \end{pmatrix} = B_{\alpha}^{-1} \begin{pmatrix} D_{-\alpha} & 0 \\ 0 & D_{\alpha} \end{pmatrix} B_{\alpha} \begin{pmatrix} \vec{E} \\ \vec{H} \end{pmatrix}. \tag{9.5}$$

This equality is of great importance because it opens the way for applying the theory of $\alpha$-holomorphic functions to time–harmonic electromagnetic field theory.

**9.5** Now we examine two interesting aspects of the interrelations of the theories under consideration. First of all, the quaternionic Cauchy–Riemann operators allow us to extract a square root of the matrix Helmholtz operator $\begin{pmatrix} \Delta_{\lambda} & 0 \\ 0 & \Delta_{\lambda} \end{pmatrix}$ which acts on $C^2(\Omega; \mathbb{H}(\mathbb{C}) \times \mathbb{H}(\mathbb{C}))$. In fact, equality (2.18) implies

$$\begin{pmatrix} \Delta_{\lambda} & 0 \\ 0 & \Delta_{\lambda} \end{pmatrix} = \begin{pmatrix} D_{-\alpha} & 0 \\ 0 & D_{\alpha} \end{pmatrix} \begin{pmatrix} -D_{\alpha} & 0 \\ 0 & -D_{-\alpha} \end{pmatrix}. \tag{9.6}$$

But for the matrix operators on the right-hand side we have:

$$\begin{pmatrix} D_\alpha & 0 \\ 0 & D_{-\alpha} \end{pmatrix} = -\begin{pmatrix} 0 & 1 \\ 1 & 0 \end{pmatrix}\begin{pmatrix} D_{-\alpha} & 0 \\ 0 & D_\alpha \end{pmatrix}\begin{pmatrix} 0 & 1 \\ 1 & 0 \end{pmatrix}.$$

Therefore we obtain:

$$\begin{pmatrix} \Delta_\lambda & 0 \\ 0 & \Delta_\lambda \end{pmatrix} = -\begin{pmatrix} D_{-\alpha} & 0 \\ 0 & D_\alpha \end{pmatrix}\begin{pmatrix} 0 & 1 \\ 1 & 0 \end{pmatrix}\begin{pmatrix} D_{-\alpha} & 0 \\ 0 & D_\alpha \end{pmatrix}\begin{pmatrix} 0 & 1 \\ 1 & 0 \end{pmatrix},$$

from which

$$\begin{pmatrix} \Delta_\lambda & 0 \\ 0 & \Delta_\lambda \end{pmatrix} = -\begin{pmatrix} 0 & D_{-\alpha} \\ D_\alpha & 0 \end{pmatrix}\begin{pmatrix} 0 & D_{-\alpha} \\ D_\alpha & 0 \end{pmatrix} =$$

$$= \begin{pmatrix} 0 & -D_{-\alpha} \\ D_\alpha & 0 \end{pmatrix}\begin{pmatrix} 0 & -D_{-\alpha} \\ D_\alpha & 0 \end{pmatrix}.$$

Finally,

$$\begin{pmatrix} \Delta_\lambda & 0 \\ 0 & \Delta_\lambda \end{pmatrix} = \begin{pmatrix} 0 & -D_{-\alpha} \\ D_\alpha & 0 \end{pmatrix}^2.$$

**9.6** The second aspect is induced by the following: there exists a close connection between generalized holomorphic vectors and $\alpha$-holomorphic functions (with $\alpha \in \mathbb{C}$, see Subsection 3.19) as well as between the latter functions and electromagnetic fields (see Subsection 9.1). Thus, it is natural to look for a precise relation between the generalized holomorphic vectors and time-harmonic electromagnetic fields.

Let us introduce the set

$$\mathcal{G} := \{(f,g) \mid f \in \mathfrak{M}_{-\vec{\alpha}}(\Omega), g \in \mathfrak{M}_{\vec{\alpha}}(\Omega), \mathrm{Sc}((f+g)\vec{\alpha}) = 0, \mathrm{Sc}(f) = \mathrm{Sc}(g)\}.$$

Then the following proposition holds.

**9.7 Proposition** *If $\vec{\alpha}^2 = i\omega\mu\sigma$ then there exists a one-to-one correspondence between $\mathcal{G}$ and the set of pairs $(\vec{E}, \vec{H})$ satisfying the Maxwell equations (8.12), given by the formulas:*

$$B_{\vec{\alpha}}(\vec{E}, \vec{H})^T := (\sigma\vec{E} + \vec{H}\vec{\alpha}, -\sigma\vec{E} + \vec{H}\vec{\alpha})^T,$$

$$B_{\vec{a}}^{-1}(f,g)^T := ((2\sigma)^{-1}(f-g),(f+g)(2\vec{\alpha})^{-1})^T.$$

*(compare with (8.5)).*

PROOF. Let the pair $(\vec{E},\vec{H})$ satisfy (8.12). Taking $(f,g)^T := B_{\vec{a}}(\vec{E},\vec{H})^T$ we have

$$D_{-\vec{a}}[f] = D_{-\vec{a}}[\sigma\vec{E} + \vec{H}\vec{\alpha}] = \sigma D[\vec{E}] - \sigma\vec{E}\vec{\alpha} + D[\vec{H}]\vec{\alpha} - \vec{H}\vec{\alpha}^2 = 0.$$

By analogy, $D_{\vec{a}}[g] = 0$. Further,

$$\mathrm{Sc}((f+g)\vec{\alpha}) = \mathrm{Sc}(2\vec{H}\vec{\alpha}^2) = 0,$$

$$\mathrm{Sc}(f) = -<\vec{H},\vec{\alpha}> = \mathrm{Sc}(g).$$

Hence, $B_{\vec{a}}$ maps every time–harmonic electromagnetic field $(\vec{E},\vec{H})$ to $\mathcal{G}$.

Now let $(f,g)\in\mathcal{G}$ and $(\vec{E},\vec{H})^T := B_{\vec{a}}^{-1}(f,g)^T$. Then we have

$$\mathrm{Sc}(\vec{E}) = \mathrm{Sc}((2\sigma)^{-1}(f-g)) = 0,$$

$$\mathrm{Sc}(\vec{H}) = \mathrm{Sc}((f+g)(2\vec{\alpha})^{-1}) = -2\mid\vec{\alpha}\mid^{-2}\mathrm{Sc}((f+g)\vec{\alpha}) = 0.$$

Consider

$$D[\vec{E}] = (2\sigma)^{-1}D[f-g] = (2\sigma)^{-1}(f+g)\vec{\alpha} = i\omega\mu\vec{H}.$$

Similarly, $D[\vec{H}] = \sigma\vec{E}$. Consequently, the pair $(\vec{E},\vec{H})$ satisfies (8.14) and, therefore, (8.12). □

# 10 Integral representations of electrodynamical quantities

**10.1** We now combine the relationships established in Section 9 and the results of Section 4. Taking into account the exceptional role of Cauchy–type integrals in $\alpha$-holomorphic function theory, we start with its analogue for time–harmonic electromagnetic fields.

Let $(\vec{E},\vec{H})$ be a time–harmonic electromagnetic field in $\Omega$, and let, as earlier,

$$\begin{pmatrix} \vec{\varphi} \\ \vec{\psi} \end{pmatrix} := B_\alpha \begin{pmatrix} \vec{E} \\ \vec{H} \end{pmatrix}.$$

As $(\vec{\varphi},\vec{\psi}) \in \mathfrak{M}_{-\alpha} \times \mathfrak{M}_\alpha$ we can apply Theorem 4.12 to the functions $\vec{\varphi}$ and $\vec{\psi}$ yielding the following chain of equalities (in the domain $\Omega$):

$$(\vec{E},\vec{H})^T = B_\alpha^{-1}(\vec{\varphi},\vec{\psi})^T =$$

$$= B_\alpha^{-1}\begin{pmatrix} K_{-\alpha} & 0 \\ 0 & K_\alpha \end{pmatrix}\begin{pmatrix} \vec{\varphi} \\ \vec{\psi} \end{pmatrix} = B_\alpha^{-1}\begin{pmatrix} K_{-\alpha} & 0 \\ 0 & K_\alpha \end{pmatrix} B_\alpha \begin{pmatrix} \vec{E} \\ \vec{H} \end{pmatrix}. \tag{10.1}$$

Thus it seems natural to call the operator

$$\mathbb{K}_\alpha := B_\alpha^{-1}\begin{pmatrix} K_{-\alpha} & 0 \\ 0 & K_\alpha \end{pmatrix} B_\alpha =$$

$$= \begin{pmatrix} \frac{1}{2}(K_{-\alpha}+K_\alpha) & \frac{\alpha}{2\sigma}(K_{-\alpha}-K_\alpha) \\ \\ \frac{\sigma}{2\alpha}(K_{-\alpha}-K_\alpha) & \frac{1}{2}(K_{-\alpha}+K_\alpha) \end{pmatrix} \tag{10.2}$$

the analogue of the Cauchy–type operator for time–harmonic electromagnetic fields. Occasionally, we shall call it "the electromagnetic Cauchy operator".

The equality (10.1) shows that the operator (10.2) acts invariantly on electromagnetic fields in the sense that it transforms an electromagnetic field into an electromagnetic field. This means, in particular, that although formally $\mathbb{K}_\alpha(\vec{E},\vec{H})^T$ is a pair of four–dimensional complex quaternions, in fact the scalar parts of both of them are equal to zero (identically on electromagnetic fields).

Let us write down explicitly in vectorial language the equality (10.1) after having separated the scalar and the vectorial parts:

$$0 = \int_\Gamma \{< \mathrm{grad}_x\theta_\alpha(x-y), [\vec{n}(y)\times\vec{E}(y)] > +$$

$$+ i\omega\mu\theta_\alpha(x-y) < \vec{n}(y), \vec{H}(y) >\} d\Gamma_y, \tag{10.3}$$

$$\vec{E}(x) = \int_\Gamma \{[\mathrm{grad}_x\theta_\alpha(x-y) \times [\vec{n}(y) \times \vec{E}(y)]] - \mathrm{grad}_x\theta_\alpha(x-y) < \vec{n}(y), \vec{E}(y) > +$$

$$+ i\omega\mu\theta_\alpha(x-y)[\vec{n}(y) \times \vec{H}(y)]\}d\Gamma_y, \tag{10.4}$$

$$0 = \int_\Gamma \{< \mathrm{grad}_x\theta_\alpha(x-y), [\vec{n}(y) \times \vec{H}(y)] > +$$

$$+ \sigma\theta_\alpha(x-y) < \vec{n}(y), \vec{E}(y) >\}d\Gamma_y, \tag{10.5}$$

$$\vec{H}(x) = \int_\Gamma \{[\mathrm{grad}_x\theta_\alpha(x-y) \times [\vec{n}(y) \times \vec{H}(y)]] - \mathrm{grad}_x\theta_\alpha(x-y) < \vec{n}(y), \vec{H}(y) > +$$

$$+ \sigma\theta_\alpha(x-y)[\vec{n}(y) \times \vec{E}(y)]\}d\Gamma_y, \quad x \in \Omega. \tag{10.6}$$

These formulas may be proved in the traditional way. For example, the scalar identities (10.3) and (10.5) are particular cases of the Gauss theorem (see Subsection 7.4). The vector equalities (10.4), (10.6) are the well-known (see, e.g., [32, p.93], [50, p.34], [136, 28, 89]) Stratton–Chu formulas. Thus, the Stratton–Chu formulas are the coordinate-wise way of writing of the quaternionic equality (10.1). In this sense they are a good analogue of the complex Cauchy integral formula.

We would like to emphasize that, in our opinion, formula (10.1) is much more transparent and heuristic than its vectorial form, system (10.4), (10.6). Moreover, the simple (after Section 9 is known) idea of how to obtain (10.1) can be developed in many directions with corresponding consequences for time–harmonic electromagnetic field theory.

Let us illustrate this by obtaining analogues of the Cauchy integral theorem and the Morera theorem.

**10.2 Theorem** (Integral form of the Maxwell equations; analogue of the Cauchy integral theorem). *Let $\Omega \in \mathbb{R}^3$ have a closed Liapunov boundary $\Gamma = \partial\Omega$. Let $(\vec{E}, \vec{H})$ be an electromagnetic time-harmonic field in $\Omega$. Then*

$$i\omega\mu\int_\Omega \vec{H}dx = \int_\Gamma [\vec{n}\times\vec{E}]d\Gamma, \qquad \int_\Gamma <\vec{n},\vec{E}> d\Gamma = 0, \tag{10.7}$$

$$\sigma\int_\Omega \vec{E}dx = \int_\Gamma [\vec{n}\times\vec{H}]d\Gamma, \qquad \int_\Gamma <\vec{n},\vec{H}> d\Gamma = 0. \tag{10.8}$$

**Proof.** Denoting as before $(\vec{\varphi},\vec{\psi})^T = B_\alpha(\vec{E},\vec{H})^T$ we have from (4.8):

$$\alpha\int_\Omega(-\sigma\vec{E}+\alpha\vec{H})dx = -\int_\Gamma \vec{n}(-\sigma\vec{E}+\alpha\vec{H})d\Gamma, \tag{10.9}$$

$$\alpha\int_\Omega(\sigma\vec{E}+\alpha\vec{H})dx = \int_\Gamma \vec{n}(\sigma\vec{E}+\alpha\vec{H})d\Gamma. \tag{10.10}$$

Adding and subtracting (10.9) and (10.10) we arrive at

$$i\omega\mu\int_\Omega \vec{H}dx = \int_\Gamma \vec{n}\vec{E}d\Gamma, \tag{10.11}$$

$$\sigma\int_\Omega \vec{E}dx = \int_\Gamma \vec{n}\vec{H}d\Gamma. \tag{10.12}$$

This is all, since (10.7) and (10.8) are the result of separating the scalar and the vector parts in (10.11) and (10.12). □

**10.3 Theorem** (Morera's theorem; condition of conversion of the integral form of the Maxwell equations) *If*

$$\{\vec{E},\vec{H}\}\subset C^1(\Omega;\mathbb{C}^3), \qquad \{D_\alpha[-\sigma\vec{E}+\alpha\vec{H}],\ \ D_{-\alpha}[\sigma\vec{E}+\alpha\vec{H}]\}\subset L_r(\Omega;\mathbb{H}(\mathbb{C}))$$

*for some $r>1$ and if for any Liapunov manifold-without-boundary $\hat{\Gamma}$ ($\hat{\Gamma}:=\partial\hat{\Omega};\hat{\Omega}\subset\Omega$) the following equalities hold:*

$$\begin{aligned} i\omega\mu\int_{\hat\Omega} \vec{H}dx &= \int_{\hat\Gamma} [\vec{n}\times\vec{E}]d\hat\Gamma, \qquad \int_{\hat\Gamma} <\vec{n},\vec{E}> d\hat\Gamma = 0, \\ \sigma\int_{\hat\Omega} \vec{E}dx &= \int_{\hat\Gamma} [\vec{n}\times\vec{H}]d\hat\Gamma, \qquad \int_{\hat\Gamma} <\vec{n},\vec{H}> d\hat\Gamma = 0, \end{aligned} \tag{10.13}$$

*then the pair $(\vec{E},\vec{H})$ is a solution to (8.12).*

**Proof.** The scalar and vector equalities (10.13) imply the quaternionic equalities

$$i\omega\mu \int_{\hat\Omega} \vec{H} dx = \int_{\hat\Gamma} \vec{n}\vec{E} d\hat\Gamma, \qquad \sigma \int_{\hat\Omega} \vec{E} dx = \int_{\hat\Gamma} \vec{n}\vec{E} d\hat\Gamma,$$

which is equivalent to the system

$$\alpha \int_{\hat\Omega} (-\sigma\vec{E} + \alpha\vec{H}) dx = -\int_{\hat\Gamma} \vec{n}(-\sigma\vec{E} + \alpha\vec{H}) d\hat\Gamma,$$

$$\alpha \int_{\hat\Omega} (\sigma\vec{E} + \alpha\vec{H}) dx = \int_{\hat\Gamma} \vec{n}(\sigma\vec{E} + \alpha\vec{H}) d\hat\Gamma.$$

Thus each of the functions $\sigma\vec{E} + \alpha\vec{H}$ and $-\sigma\vec{E} + \alpha\vec{H}$ satisfies the quaternionic Morera theorem (Theorem 4.13). Hence, $\sigma\vec{E} + \alpha\vec{H} \in \mathfrak{M}_{-\alpha}$, $-\sigma\vec{E} + \alpha\vec{H} \in \mathfrak{M}_{\alpha}$. The latter is equivalent to the conclusion of the theorem. □

**10.4 Theorem** (Quaternionic Stokes formula and the Umov–Poynting theorem) *For any electromagnetic time-harmonic field, the following equalities hold:*

$$\int_{\Gamma} < \vec{n}, [\vec{E} \times \vec{H}] > d\Gamma = i\omega\mu \int_{\Omega} < \vec{H}, \vec{H} > dx - \sigma \int_{\Omega} < \vec{E}, \vec{E} > dx;$$

$$\int_{\Gamma} (< \vec{n}, \vec{H} > \vec{E} + < \vec{n}, \vec{E} > \vec{H}) d\Gamma = 0. \tag{10.14}$$

**Proof.** Substitute $\vec{E}$ for $g$ and $\vec{H}$ for $f$ into (4.3), and take into account $D_r[\vec{E}] = -D[\vec{E}]$. Then we have:

$$\int_{\Gamma} (\vec{E}\vec{n}\vec{H}) d\Gamma = \int_{\Omega} (-D[\vec{E}]\vec{H} + \vec{E}D[\vec{H}]) dx.$$

By the properties of solutions of the Maxwell equations the latter gives:

$$\int_{\Gamma} (\vec{E}\vec{n}\vec{H}) d\Gamma = \int_{\Omega} (-i\omega\mu\vec{H}^2 + \sigma\vec{E}^2) dx. \tag{10.15}$$

Now rewrite the equality (10.15) in vectorial language:

$$\int_{\Gamma} [\vec{E} \times [\vec{n} \times \vec{H}]] d\Gamma - \int_{\Gamma} \vec{E} < \vec{n}, \vec{H} > d\Gamma - \int_{\Gamma} < \vec{E}, [\vec{n} \times \vec{H}] > d\Gamma =$$

$$= -\sigma \int_{\Omega} < \vec{E}, \vec{E} > dx + i\omega\mu \int_{\Omega} < \vec{H}, \vec{H} > dx. \tag{10.16}$$

Using the vectorial identity

$$[\vec{E} \times [\vec{n} \times \vec{H}]] = \vec{n} < \vec{E}, \vec{H} > - \vec{H} < \vec{E}, \vec{n} >$$

and the property of an electromagnetic field $< \vec{E}, \vec{H} >= 0$ (see, e.g., [50, p.12]) we have

$$\int_{\Gamma} \vec{H} < \vec{E}, \vec{n} > d\Gamma + \int_{\Gamma} \vec{E} < \vec{n}, \vec{H} > d\Gamma + \int_{\Gamma} < \vec{E}, [\vec{n} \times \vec{H}] > d\Gamma =$$

$$= \sigma \int_{\Omega} < \vec{E}, \vec{E} > dx - i\omega\mu \int_{\Omega} < \vec{H}, \vec{H} > dx. \tag{10.17}$$

Now (10.14) is nothing but the equalities of the scalar parts and of the vectorial parts in (10.17). □

The first equality in (10.14) is known as the Umov–Poynting theorem (see, e.g., [92, p.116]).

**10.5** Let $(\vec{E}, \vec{H})$ be an electromagnetic field in $\Omega$. It is well-known (see [2, p.95]) that its impulse is described by the integral

$$\int_{\Omega} [\vec{E} \times \vec{H}] dx.$$

The following theorem gives a representation of this integral in the form of an integral over the boundary of $\Omega$.

**10.6 Theorem** (Impulse of an electromagnetic field as the integral over the boundary) *Let $(\vec{E}, \vec{H})$ be an electromagnetic time-harmonic field in $\Omega$. Then*

$$4\alpha^2 \int_{\Omega} [\vec{E} \times \vec{H}] dx = \int_{\Gamma} (-2\sigma \vec{E} < \vec{n}, \vec{E} > +$$

$$+ \sigma \vec{n} < \vec{E}, \vec{E} > + 2i\omega\mu \vec{H} < \vec{n}, \vec{H} > - i\omega\mu \vec{n} < \vec{H}, \vec{H} >) d\Gamma. \tag{10.18}$$

PROOF. Apply to the pair of functions

$$(\vec{\varphi}, \vec{\psi})^T := B_\alpha(\vec{E}, \vec{H})^T$$

the quaternionic Stokes formula (4.3):

$$\int_\Gamma \vec{\varphi}\vec{n}\vec{\psi} d\Gamma = -2\alpha \int_\Omega \vec{\varphi}\vec{\psi} dx.$$

The result is

$$\int_\Gamma (\sigma\vec{E} + \alpha\vec{H})\vec{n}(-\sigma\vec{E} + \alpha\vec{H}) d\Gamma = -2\alpha \int_\Omega (\sigma\vec{E} + \alpha\vec{H})(-\sigma\vec{E} + \alpha\vec{H}) dx,$$

which is equivalent to the equality

$$\int_\Gamma (-\sigma^2\vec{E}\vec{n}\vec{E} + \alpha^2\vec{H}\vec{n}\vec{H} + \alpha\sigma(\vec{E}\vec{n}\vec{H} - \vec{H}\vec{n}\vec{E})) d\Gamma =$$

$$= 2\alpha \int_\Omega (\sigma^2\vec{E}^2 - \alpha^2\vec{H}^2 - 2\alpha\sigma\vec{E}\vec{H}) dx. \tag{10.19}$$

Using (10.15) we have

$$\int_\Gamma (\vec{E}\vec{n}\vec{H} - \vec{H}\vec{n}\vec{E}) d\Gamma = 2 \int_\Omega (\sigma\vec{E}^2 - i\omega\mu\vec{H}^2) dx.$$

Then the equality (10.19) turns into

$$4\alpha^2 \int_\Omega \vec{E}\vec{H} dx = \int_\Gamma (\sigma\vec{E}\vec{n}\vec{E} - i\omega\mu\vec{H}\vec{n}\vec{H}) d\Gamma. \tag{10.20}$$

Rewriting (10.20) in vectorial form we obtain (10.18). □

# 11 Boundary value problems for time-harmonic electromagnetic fields

**11.1** In Section 6 some boundary value problems for $\alpha$-holomorphic functions were considered. In Section 9 the relations between time-harmonic electromagnetic fields and $\alpha$-holomorphic functions were established. This allows us to use the results of Section 6 for solving some boundary value problems for time-harmonic electromagnetic fields.

We start with the following

**11.2 Problem** Given a pair of functions $\{\vec{e}, \vec{h}\} : \Gamma \to \mathbb{C}^3$; to find a pair of functions $\{\vec{E}, \vec{H}\} : \bar{\Omega} \to \mathbb{C}^3$ which satisfy the Maxwell equations (8.12) in $\Omega$ and the equalities

$$\vec{E}|_\Gamma = \vec{e}, \quad \vec{H}|_\Gamma = \vec{h} \tag{11.1}$$

on $\Gamma$.

Its solution is described by the following result.

**11.3 Theorem** *Given a closed Liapunov surface $\Gamma := \partial\Omega \subset \mathbb{R}^3$ and functions $\vec{e}, \vec{h} \in C^{0,\epsilon}(\Gamma; \mathbb{R}^3)$, they are boundary values of solutions to the Maxwell equations (8.12) in $\Omega$ if, and only if, the following equalities are satisfied:*

$$\begin{aligned} \vec{e}(\tau) &= 2\int_\Gamma \{[\operatorname{grad}_\tau \theta_\alpha(\tau - y) \times [\vec{n}(y) \times \vec{e}(y)]] - \\ &- \operatorname{grad}_\tau \theta_\alpha(\tau - y) < \vec{n}(y), \vec{e}(y) > + \\ &+ i\omega\mu\theta_\alpha(\tau - y)[\vec{n}(y) \times \vec{h}(y)]\} d\Gamma_y, \end{aligned} \tag{11.2}$$

$$\begin{aligned} \vec{h}(\tau) &= 2\int_\Gamma \{[\operatorname{grad}_\tau \theta_\alpha(\tau - y) \times [\vec{n}(y) \times \vec{h}(y)]] - \\ &- \operatorname{grad}_\tau \theta_\alpha(\tau - y) < \vec{n}(y), \vec{e}(y) > + \\ &+ \sigma\theta_\alpha(\tau - y)[\vec{n}(y) \times \vec{e}(y)]\} d\Gamma_y, \end{aligned} \tag{11.3}$$

$$\int_\Gamma \{< \operatorname{grad}_\tau \theta_\alpha(\tau - y), [\vec{n}(y) \times \vec{e}(y)] > +$$

$$+ i\omega\mu\theta_\alpha(\tau - y) < \vec{n}(y), \vec{h}(y) > \} d\Gamma_y = 0, \tag{11.4}$$

$$\int_\Gamma \{< \operatorname{grad}_\tau \theta_\alpha(\tau - y), [\vec{n}(y) \times \vec{h}(y)] > +$$

$$+ \sigma\theta_\alpha(\tau - y) < \vec{n}(y), \vec{e}(y) > \} d\Gamma_y = 0, \tag{11.5}$$

*for any* $\tau \in \Gamma$, *where the integrals exist in the sense of Cauchy's principal value;* $\alpha \in \mathbb{C}$, $\alpha^2 = i\omega\mu\sigma$.

PROOF. Let $\vec{e}, \vec{h}$ satisfying the Hölder condition be the boundary values of the solutions to (8.12). That is, there exist $\vec{E}, \vec{H}$ satisfying (8.12) as well as (11.1). Then in $\Omega$

$$\vec{E} = \frac{1}{2\sigma}\{K_{-\alpha}[\sigma\vec{e} + \alpha\vec{h}] - K_\alpha[-\sigma\vec{e} + \alpha\vec{h}]\}, \tag{11.6}$$

$$\vec{H} = \frac{1}{2\alpha}\{K_{-\alpha}[\sigma\vec{e} + \alpha\vec{h}] + K_\alpha[-\sigma\vec{e} + \alpha\vec{h}]\} \tag{11.7}$$

(see formula (10.1)).

Both sides of (11.6) and (11.7) are functions of $x \in \Omega$. Now let $x \to \tau \in \Gamma$. Using the quaternionic Plemelj–Sokhotski formula we obtain on $\Gamma$:

$$\vec{e} = \frac{1}{2\sigma}\{P_{-\alpha}[\sigma\vec{e} + \alpha\vec{h}] - P_\alpha[-\sigma\vec{e} + \alpha\vec{h}]\},$$

$$\vec{h} = \frac{1}{2\alpha}\{P_{-\alpha}[\sigma\vec{e} + \alpha\vec{h}] + P_\alpha[-\sigma\vec{e} + \alpha\vec{h}]\}.$$

Consequently on $\Gamma$ we have

$$\vec{e} = \frac{1}{2\sigma}\{S_{-\alpha}[\sigma\vec{e} + \alpha\vec{h}] - S_\alpha[-\sigma\vec{e} + \alpha\vec{h}]\}, \tag{11.8}$$

$$\vec{h} = \frac{1}{2\alpha}\{S_{-\alpha}[\sigma\vec{e} + \alpha\vec{h}] + S_{\alpha}[-\sigma\vec{e} + \alpha\vec{h}]\}. \tag{11.9}$$

Rewriting (11.8), (11.9) in vectorial form and separating the scalar and the vector parts we obtain (11.2)–(11.5).

Let, on the contrary, the equalities (11.2)–(11.5) be satisfied. Then (11.8), (11.9) are valid. Let us consider the functions $\vec{E}, \vec{H}$ chosen to be of the form (11.6), (11.7). Then in $\Omega$

$$\begin{aligned} D[\vec{E}] &= \frac{1}{2\sigma}(DK_{-\alpha}[\sigma\vec{e} + \alpha\vec{h}] - DK_{\alpha}[-\sigma\vec{e} + \alpha\vec{h}]) = \\ &= \frac{1}{2\sigma}(\alpha K_{-\alpha}[\sigma\vec{e} + \alpha\vec{h}] + \alpha K_{\alpha}[-\sigma\vec{e} + \alpha\vec{h}]) = i\omega\mu\vec{H} \end{aligned}$$

(we take into account that $\alpha^2 = i\omega\mu\sigma$).

By analogy, $D[\vec{H}] = \sigma\vec{E}$ in $\Omega$. Then $(\vec{E}, \vec{H})$ is a solution to (8.12). It remains only to verify (11.1). Again sending $x \in \Omega$ to $\tau \in \Gamma$ we can use the quaternionic Plemelj–Sokhotski formula (5.2). Then

$$\vec{E}|_{\Gamma} = \frac{1}{2}\{\vec{e} + \frac{1}{2\sigma}(S_{-\alpha}[\sigma\vec{e} + \alpha\vec{h}] - S_{\alpha}[-\sigma\vec{e} + \alpha\vec{h}])\}|_{\Gamma},$$

$$\vec{H}|_{\Gamma} = \frac{1}{2}\{\vec{h} + \frac{1}{2\alpha}(S_{-\alpha}[\sigma\vec{e} + \alpha\vec{h}] + S_{\alpha}[-\sigma\vec{e} + \alpha\vec{h}])\}|_{\Gamma}.$$

Then (11.1) follows from (11.8), (11.9).

□

**11.4 Remark** A similar result was shown in [135, p. 259] (see also [136]). But instead of the scalar integral conditions (11.4), (11.5) the following conditions were obtained there:

$$< \vec{e}, \vec{n} >= -\frac{1}{\sigma}Div[\vec{n} \times \vec{h}],$$
$$< \vec{h}, \vec{n} >= -\frac{1}{i\omega\mu}Div[\vec{n} \times \vec{e}],$$

where $Div$ denotes the surface divergence, hence an additional restrictive condition of the differentiability of tangential components of $\vec{e}, \vec{h}$ on $\Gamma$ was necessary.

**11.5 Remark** From Theorem 11.3 it follows that for the functions $\vec{e}, \vec{h}$ satisfying the Hölder condition the problem (11.1) is solvable if, and only if, the equalities (11.2)–(11.5) are true. Moreover, if (11.2)–(11.5) are satisfied then the solution $(\vec{E}, \vec{H})$ of the problem (11.1) is constructed by formulas (11.6), (11.7) and is unique.

Let us examine the following

**11.6 Problem** Let $\Omega = \mathbb{R}_3^+ = \{x \in \mathbb{R}^3 | x_3 > 0\}$, $\Omega := \partial\Omega = \mathbb{R}^2$ with the outward normal $\vec{n} \equiv -i_3$.

Given functions $e_3, h_3 : \Gamma \to \mathbb{C}$, to find functions $e_1, e_2, h_1, h_2 : \Gamma \to \mathbb{C}$ such that $\vec{e} := (e_1, e_2, e_3)^T$, $\vec{h} := (h_1, h_2, h_3)^T$ can be extended onto the upper half-space $\Omega$ to be an electromagnetic field there, i.e., to satisfy (8.12) in $\Omega$. We assume that $e_3, h_3$ belong to the Lizorkin space $\mathcal{L}(\Gamma)$ (see Subsection 5.24).

**11.7** Using the bijection $B_\alpha$ from Subsection 9.4 we see that Problem 11.6 is equivalent to the following quaternionic one.

Given functions $\varphi_0 = \psi_0 \equiv 0$ on $\mathbb{R}^2$, $\{\varphi_3, \psi_3\} \subset \mathcal{L}(\mathbb{R}^2)$, to find functions $\varphi_1$, $\varphi_2$, $\psi_1$, $\psi_2$ such that $\vec{\varphi} := \sum_{k=1}^3 \varphi_k i_k$ is $(-\alpha)$-holomorphically extendable onto $\Omega$, and $\vec{\psi}$ is $\alpha$-holomorphically extendable onto $\Omega$.

Let us show that in fact the extensions of $\vec{\varphi}$ and $\vec{\psi}$ will be purely vectorial also. This follows from simple reasoning. Every component of a $(\pm\alpha)$-holomorphic function is metaharmonic in $\Omega$:

$$\Delta\varphi_k + \alpha^2\varphi_k = 0. \tag{11.10}$$

But the exterior Dirichlet problem for the Helmholtz equation is uniquely solvable (see, e.g., [27, p. 83]), thus the conditions $\varphi_0 = \psi_0 = 0$ on $\mathbb{R}^2$ imply that their extensions are identically zero on $\mathbb{R}_+^3$. Hence Problem 11.6 reduces to the pair of problems with the boundary conditions

$$\varphi_0|_{\mathbb{R}^2} = 0, \quad \varphi_3|_{\mathbb{R}^2} = b_1, \tag{11.11}$$

and

$$\psi_0|_{\mathbb{R}^2} = 0, \quad \psi_3|_{\mathbb{R}^2} = b_2, \tag{11.12}$$

where $b_1, b_2$ are given functions from the Lizorkin space; $b_1 := \sigma e_3 + \alpha h_3, b_2 := -\sigma e_3 + \alpha h_3$.

We can write down the solutions $\varphi, \psi$ immediately using Subsection 6.6:

$$\varphi = K_{-\alpha}[f],$$

where the first bicomplex component of $f$ is given by $F_1 = b_1 \boldsymbol{i}_3$ and the second one by $F_2 = H_{-\alpha}^{-1}[F_1]$ with $H_{-\alpha}^{-1}$ defined in Subsection 5.27.

By analogy,

$$\psi = K_\alpha[g],$$

where $G_1 = b_2 \boldsymbol{i}_3$, $G_2 = H_\alpha^{-1}[G_1]$.

Then the required solution $(\vec{E}, \vec{H})$ of (8.12) in $\Omega$ has the form

$$\begin{aligned} \vec{E} &= \frac{1}{2\sigma}(K_{-\alpha}[H_{-\alpha}^{-1}[b_1 \boldsymbol{i}_3]\boldsymbol{i}_1 + b_1 \boldsymbol{i}_3] - \\ &- K_\alpha[H_\alpha^{-1}[b_2 \boldsymbol{i}_3]\boldsymbol{i}_1 + b_2 \boldsymbol{i}_3]), \end{aligned} \tag{11.13}$$

$$\begin{aligned} \vec{H} &= \frac{1}{2\alpha}(K_{-\alpha}[H_{-\alpha}^{-1}[b_1 \boldsymbol{i}_3]\boldsymbol{i}_1 + b_1 \boldsymbol{i}_3] + \\ &+ K_\alpha[H_\alpha^{-1}[b_2 \boldsymbol{i}_3]\boldsymbol{i}_1 + b_2 \boldsymbol{i}_3]). \end{aligned} \tag{11.14}$$

We used only the map $B_\alpha^{-1}$ (see Subsection 9.4).

Thus, Problem 11.6 is uniquely solvable and its solution is given by (11.13)–(11.14). We immediately obtain the following

**11.8 Theorem** (Hilbert formulas for time-harmonic electromagnetic fields) *Given two complex-valued functions* $e_3, h_3$ *on* $\mathbb{R}^2 = \Gamma := \{x \in \mathbb{R}^3 | x_3 = 0, \vec{n} \equiv -i_3\}$, $\{e_3, h_3\} \subset \mathcal{L}(\mathbb{R}^2)$ . *Then the pair* $(\vec{e}, \vec{h})$ *of* $\mathbb{C}^3$ *-valued functions on* $\mathbb{R}^2$ *such that it is a boundary value of a time-harmonic electromagnetic field in* $\mathbb{R}^+_3 := \{x \in \mathbb{R}^3 | x_3 > 0\}$ *is unique, and the components* $e_1, e_2, h_1, h_2$ *are given by the formulas*

$$\begin{aligned} e_1 i_1 + e_2 i_2 &= \frac{1}{2\sigma}(H^{-1}_{-\alpha}[(\sigma e_3 + \alpha h_3) i_3] i_1 - \\ &- H^{-1}_{\alpha}[(-\sigma e_3 + \alpha h_3) i_3] i_1), \\ h_1 i_1 + h_2 i_2 &= \frac{1}{2\alpha}(H^{-1}_{-\alpha}[(\sigma e_3 + \alpha h_3) i_3] i_1 + \\ &+ H^{-1}_{\alpha}[(-\sigma e_3 + \alpha h_3) i_3] i_1), \end{aligned}$$

*where the operators* $H^{-1}_{\pm\alpha}$ *are defined in Subsection 5.27.*

**11.9 Corollary** (Reconstruction of a time-harmonic electromagnetic field from the normal component of its boundary value) *Under the conditions of Theorem 11.8, the field* $(\vec{E}, \vec{H})$ *in* $\mathbb{R}^+_3$ *is determined uniquely by the equalities:*

$$\begin{aligned} \vec{E} &= \frac{1}{2\sigma}(K_{-\alpha}[H^{-1}_{-\alpha}[(\sigma e_3 + \alpha h_3) i_3] i_1 + (\sigma e_3 + \alpha h_3) i_3] - \\ &- K_{\alpha}(H^{-1}_{\alpha}[(-\sigma e_3 + \alpha h_3) i_3] i_1 + (-\sigma e_3 + \alpha h_3) i_3]), \\ \vec{H} &= \frac{1}{2\alpha}(K_{-\alpha}[H^{-1}_{-\alpha}[(\sigma e_3 + \alpha h_3) i_3] i_1 + (\sigma e_3 + \alpha h_3) i_3] + \\ &+ K_{\alpha}(H^{-1}_{\alpha}[(-\sigma e_3 + \alpha h_3) i_3] i_1 + (-\sigma e_3 + \alpha h_3) i_3]). \end{aligned}$$

**11.10** Up to now we have applied the results of $\alpha$-holomorphic function theory to the solutions of the Maxwell equations. It is clear that having the bijection $B_\alpha$ we can apply known results from electromagnetic field theory to $\alpha$-holomorphic functions. Here we will give one example of such an application concerning generalized holomorphic vectors (Subsection 3.11).

We will use the one-to-one correspondence between the set

$\mathcal{G} := \{(f,g) | f \in \mathfrak{M}_{-\vec{\alpha}}(\Omega), g \in \mathfrak{M}_{\vec{\alpha}}(\Omega), \mathrm{Sc}((f+g)\vec{\alpha}) = 0\,,\ \mathrm{Sc}(f) = \mathrm{Sc}(g)\}$

of generalized holomorphic vectors and the set of time-harmonic electromagnetic fields established by Proposition 9.7.

**11.11 Problem** To find a pair $(f,g) \in \mathcal{G} \cap C(\bar{\Omega})$ satisfying the boundary condition

$$[\vec{n} \times (f-g)]|_\Gamma = \vec{c}|_\Gamma, \tag{11.15}$$

where $\Gamma$ is a closed Liapunov surface in $\mathbb{R}^3, \Gamma := \partial\Omega\,;\ \vec{c} \in C^{0,\epsilon}(\Gamma)$ is a given vector field with an additional condition: Div $\vec{c} \in C^{0,\epsilon}(\Gamma)$.

**11.12** From Proposition 9.7 it follows that $(f,g)$ solves Problem 11.11 iff the pair $(\vec{E},\vec{H}) := B_{\vec{\alpha}}^{-1}(f,g)$ satisfies (8.12) with the boundary condition

$$[\vec{n} \times \vec{E}]|_\Gamma = \frac{1}{2\sigma}\vec{c}|_\Gamma. \tag{11.16}$$

The interior Maxwell boundary value problem (8.12), (11.16) is considered in [27, p. 121].

Using the corresponding solvability criterion [27, Theorem 4.26] we obtain immediately the following

**11.13 Theorem** *Problem 11.11 is solvable iff the correlation*

$$\int_\Gamma < \vec{c}, (f+g)\vec{\alpha} > d\Gamma = 0$$

*holds for all solutions $f,g$ of the homogeneous problem (11.15).*

# Chapter 3

# Massive spinor fields

## 12 The classical Dirac equation and its biquaternionic reformulation

**12.1** It was established in Subsection 1.4 that the map $\kappa : \mathbb{H}(\mathbb{C}) \longrightarrow \mathfrak{D}$ gives an isomorphism of the algebra $\mathbb{H}(\mathbb{C})$ onto a subalgebra of the Dirac algebra.

Now for any $A \in \mathfrak{D}$, that is, for any matrix $A$ of the form (1.14) we define a column-matrix $\chi(A)$ by the rule:

$$\chi(A) := A \begin{pmatrix} 1 \\ 0 \\ 0 \\ 0 \end{pmatrix}.$$

Thus we have a map

$$\chi : A \in \mathfrak{D} \longmapsto \beta := \chi(A) := \begin{pmatrix} \tilde{a}_0 - i\tilde{a}_3 \\ i\tilde{a}_1 - \tilde{a}_2 \\ i\tilde{\tilde{a}}_0 + \tilde{\tilde{a}}_3 \\ -\tilde{\tilde{a}}_1 - i\tilde{\tilde{a}}_2 \end{pmatrix}$$

which assigns to any matrix $A \in \mathfrak{D}$ its first column.

The map $\chi$ is invertible since by a column of complex numbers one can uniquely reconstruct a matrix of the form (1.14):

$$A := \chi^{-1}(\beta) = \chi^{-1} \begin{pmatrix} \beta_0 \\ \beta_1 \\ \beta_2 \\ \beta_3 \end{pmatrix},$$

where the real numbers $\tilde{a}_k, \tilde{\tilde{a}}_k, k \in \mathbb{N}_3^0$ in (1.13) are defined as follows:

$$\tilde{a}_0 = \mathrm{Re}\beta_0, \quad \tilde{a}_1 = \mathrm{Im}\beta_1, \quad \tilde{a}_2 = -\mathrm{Re}\beta_1, \quad \tilde{a}_3 = -\mathrm{Im}\beta_0,$$

$$(12.1)$$

$$\tilde{\tilde{a}}_0 = \mathrm{Im}\beta_2, \quad \tilde{\tilde{a}}_1 = -\mathrm{Re}\beta_3, \quad \tilde{\tilde{a}}_2 = -\mathrm{Im}\beta_3, \quad \tilde{\tilde{a}}_3 = \mathrm{Re}\beta_2.$$

From now on let us adopt the following agreement: for any domain $\Xi \subset \mathbb{R}^4$ we denote by $\tilde{\Xi}$ its image under the reflection $x_3 \mapsto -x_3$, and for any $\Phi$ defined on $\tilde{\Xi}$ the function $\tilde{\Phi}$ acts on $\Xi$ by the rule:

$$\tilde{\Phi}(t, x_1, x_2, x_3) := \Phi(t, x_1, x_2, -x_3).$$

Now we introduce a map $\mathcal{A}$ by the rule: if $\Phi : \tilde{\Xi} \subset \mathbb{R}^4 \longrightarrow \mathbb{C}^4$ then

$$F := \mathcal{A}[\Phi] := \chi^{-1}\kappa^{-1}[\tilde{\Phi}] : \Xi \subset \mathbb{R}^4 \longrightarrow \mathbb{H}(\mathbb{C}).$$

In explicit form,

$$\mathcal{A}[\Phi] := (\mathrm{Re}\tilde{\Phi}_0 + i\mathrm{Im}\tilde{\Phi}_2)i_0 + (\mathrm{Im}\tilde{\Phi}_1 - i\mathrm{Re}\tilde{\Phi}_3)i_1 -$$

$$(12.2)$$

$$(\mathrm{Re}\tilde{\Phi}_1 + i\mathrm{Im}\tilde{\Phi}_3)i_2 - (\mathrm{Im}\tilde{\Phi}_0 - i\mathrm{Re}\tilde{\Phi}_2)i_3.$$

Then we have

$$\mathcal{A}^{-1}[F] = (\mathrm{Re}\tilde{F}_0 - i\mathrm{Re}\tilde{F}_3, -\mathrm{Re}\tilde{F}_2 + i\mathrm{Re}\tilde{F}_1,$$

$$(12.3)$$

$$\mathrm{Im}\tilde{F}_3 + i\mathrm{Im}\tilde{F}_0, -\mathrm{Im}\tilde{F}_1 - i\mathrm{Im}\tilde{F}_2)^T.$$

Now we give some simple but important arithmetical properties of $\mathcal{A}$.

**12.2 Lemma** *Let* $\Phi : \tilde{\Xi} \subset \mathbb{R}^4 \longrightarrow \mathbb{C}^4$. *Then*

*1.* $\mathcal{A}[\gamma_0\Phi] = (\mathcal{A}[\Phi])^*$;

*2.* $\mathcal{A}[\gamma_0\gamma_1\Phi] = i i_1 \mathcal{A}[\Phi]$;

*3.* $\mathcal{A}[\gamma_0\gamma_2\Phi] = i i_2 \mathcal{A}[\Phi]$;

*4.* $\mathcal{A}[\gamma_0\gamma_3\Phi] = -i i_3 \mathcal{A}[\Phi]$;

*5.* $\mathcal{A}[i\Phi] = -\mathcal{A}[\Phi] i_3$;

*6.* $\mathcal{A}[\hat{i}\Phi] = i\mathcal{A}[\Phi]$.

PROOF.

1. Using (12.2) we have:

$$\mathcal{A}[\gamma_0\Phi] = \mathcal{A}\left[\begin{pmatrix} \Phi_0 \\ \Phi_1 \\ -\Phi_2 \\ -\Phi_3 \end{pmatrix}\right] =$$

$$= (\mathrm{Re}\bar{\Phi}_0 - i\mathrm{Im}\bar{\Phi}_2)i_0 + (\mathrm{Im}\bar{\Phi}_1 + i\mathrm{Re}\bar{\Phi}_3)i_1 -$$

$$(-\mathrm{Re}\bar{\Phi}_1 + i\mathrm{Im}\bar{\Phi}_3)i_2 - (\mathrm{Im}\bar{\Phi}_0 + i\mathrm{Re}\bar{\Phi}_2)i_3 = (\mathcal{A}[\Phi])^*.$$

2. Again using (12.2) we obtain:

$$\mathcal{A}[\gamma_0\gamma_1\Phi] = -\mathcal{A}\left[\begin{pmatrix} \Phi_3 \\ \Phi_2 \\ \Phi_1 \\ \Phi_0 \end{pmatrix}\right] =$$

$$= -((\mathrm{Re}\bar{\Phi}_3 + i\mathrm{Im}\bar{\Phi}_1)i_0 + (\mathrm{Im}\bar{\Phi}_2 - i\mathrm{Re}\bar{\Phi}_0)i_1 +$$

$$(-\mathrm{Re}\bar{\Phi}_2 - i\mathrm{Im}\bar{\Phi}_0)i_2 + (-\mathrm{Im}\bar{\Phi}_3 + i\mathrm{Re}\bar{\Phi}_1)i_3) = i i_1 \mathcal{A}[\Phi].$$

3.–4. The proof of these cases is completely analogous to that of 2.

5. Let us calculate

$$\mathcal{A}[i\Phi] = \mathcal{A}[i\mathrm{Re}\Phi - \mathrm{Im}\Phi] =$$

$$= \quad (-\mathrm{Im}\bar{\Phi}_0 + i\mathrm{Re}\bar{\Phi}_2)i_0 + (\mathrm{Re}\bar{\Phi}_1 + i\mathrm{Im}\bar{\Phi}_3)i_1 +$$

$$(\mathrm{Im}\bar{\Phi}_1 - i\mathrm{Re}\bar{\Phi}_3)i_2 + (-\mathrm{Re}\bar{\Phi}_0 - i\mathrm{Im}\bar{\Phi}_2)i_3) = -\mathcal{A}[\Phi]i_3.$$

6. By analogy,

$$\mathcal{A}[\hat{i}\Phi] = \kappa^{-1}\chi^{-1}\begin{pmatrix} i\Phi_2 \\ i\Phi_3 \\ i\Phi_0 \\ i\Phi_1 \end{pmatrix} =$$

$$= \kappa^{-1}(\hat{i}\chi^{-1}[\Phi]) = i\kappa^{-1}[\chi^{-1}[\Phi]] = i\mathcal{A}[\Phi].$$

□

**12.3** Let us consider the classical Dirac equation describing a free particle with spin $\frac{1}{2}$:

$$\mathbf{D}[\Phi] := (\gamma_0\partial_t - \sum_{k=1}^{3}\gamma_k\partial_k + im)[\Phi] = 0, \tag{12.4}$$

where $\partial_t := \frac{\partial}{\partial t}$ , $\Phi : \mathbb{R}^4 \to \mathbb{C}^4, m \in \mathbb{R}$ ; the standard Dirac matrices $\gamma_k$ are defined in Subsection 1.4 (the speed of light and Plank's constant divided by $2\pi$ are assumed to have the unit value, $c = \hbar = 1$ ).

Let us multiply (12.4) by the matrix $\gamma_0$ on the left-hand side and then apply the map $\mathcal{A}$:

$$\mathcal{A}[\partial_t\Phi - \sum_{k=1}^{3}\gamma_0\gamma_k\partial_k\Phi + im\gamma_0\Phi] = 0. \tag{12.5}$$

We write $F := \mathcal{A}[\Phi]$ and use Lemma 12.2; then from (12.5) we obtain

$$\partial_t F - iDF - mF^* i_3 = 0, \tag{12.6}$$

or after multiplication by $i$:

$$(i\partial_t + D - miZ_{\mathbb{C}}M^{i_3})[F] = 0. \tag{12.7}$$

Now we have the following

**12.4 Theorem** *There exists a one-to-one correspondence between the set of Dirac bispinors (i.e., the set of solutions of the Dirac equation (12.4)) and the set of solutions of the complex–quaternionic equation (12.7), which is given by the map $\mathcal{A}$: for any solution $\Phi$ of (12.4) the function $F := \mathcal{A}[\Phi]$ is a solution to (12.7), and, vice versa, for any solution $F$ to (12.7) the function $\Phi := \mathcal{A}^{-1}[F]$ is a solution to (12.4).*

PROOF. The statement follows since all transformations of Subsection 12.3 are invertible. □

**12.5** Let us denote the operator of the equation (12.7) by $N$:

$$N := i\partial_t + D - miZ_{\mathbb{C}}M^{i_3}.$$

We will consider it on continuously differentiable $\mathbb{H}(\mathbb{C})$-valued functions. Note that from the above considerations we can write

$$N = i\mathcal{A}\gamma_0\mathbb{D}\mathcal{A}^{-1}. \tag{12.8}$$

When $m = 0$, that is, the Dirac equation (12.4) describes a spinor field corresponding to a neutrino (see, e.g., [7, p.80]; [16, p.140]) the equivalent complex quaternionic equation takes the form (8.2). Hence we have an intimate connection between the equation describing the neutrino and the Maxwell equations in vacuum.

**12.6** The operator $N$ is not convenient for further investigation because of the presence of the complex conjugation operator $Z_{\mathbb{C}}$. The most urgent problem is to eliminate it.

The first step is based on the well-known matrix equality (see, e.g., [78, p.398], [73, p.88]):

$$\begin{pmatrix} I & C \\ I & -C \end{pmatrix} \begin{pmatrix} A & B \\ CBC & CAC \end{pmatrix} \begin{pmatrix} I & I \\ C & -C \end{pmatrix} = 2 \begin{pmatrix} A+BC & 0 \\ 0 & A-BC \end{pmatrix},$$

where $A, B, C$ are linear operators, $C^2 = I$. Taking $A = i\partial_t + D; B = -miM^{i_3}; C = Z_{\mathbb{C}}$ and writing $N' := i\partial_t + D + miZ_{\mathbb{C}}M^{i_3}$, we have

$$\begin{pmatrix} N & 0 \\ 0 & N' \end{pmatrix} = \frac{1}{2} \begin{pmatrix} I & Z_{\mathbb{C}} \\ I & -Z_{\mathbb{C}} \end{pmatrix} \begin{pmatrix} i\partial_t + D & -miM^{i_3} \\ miM^{i_3} & -i\partial_t + D \end{pmatrix} \begin{pmatrix} I & I \\ Z_{\mathbb{C}} & -Z_{\mathbb{C}} \end{pmatrix}.$$

That is, if $F \in \ker N, G \in \ker N'$, then the functions $f := \frac{1}{2}(F + G)$, $g := \frac{1}{2}(F^* - G^*)$ satisfy the system

$$(i\partial_t + D)f - migi_3 = 0, \tag{12.9}$$

$$(-i\partial_t + D)g + mifi_3 = 0. \tag{12.10}$$

Note that $F = f + g^*, G = f - g^*$.

To carry out the second step we introduce the notation

$$P_k^{\pm} := \frac{1}{2}M^{(1 \pm ii_k)}, \ \ k \in \mathbb{N}_3.$$

Any pair of operators $P_k^+, P_k^-$ with a fixed $k$ is a pair of mutually complementary projectors on the set of $\mathbb{H}(\mathbb{C})$-valued functions.

Let us apply $P_1^+$ to the equality (12.9) and $P_1^-$ to the equality (12.10). Then, taking into account that

$$P_1^{\pm} M^{i_3} = M^{i_3} P_1^{\mp}, \tag{12.11}$$

we obtain

$$(i\partial_t + D)P_1^+[f] - miM^{i_3}P_1^-[g] = 0, \tag{12.12}$$

$$(i\partial_t - D)P_1^-[g] - miM^{i_3}P_1^+[f] = 0. \tag{12.13}$$

Adding and subtracting the equalities (12.12), (12.13), for the functions $\varphi := P_1^+[f] - P_1^-[g]$, $\psi := P_1^-[f] - P_1^+[g]$ we have two separate equations

$$P_1^+(i\partial_t + D)[\varphi] + P_1^-(-i\partial_t + D)[\varphi] - m\varphi i_2 = 0, \tag{12.14}$$

$$P_1^+(-i\partial_t + D)[\psi] + P_1^-(i\partial_t + D)[\psi] + m\psi i_2 = 0. \tag{12.15}$$

Note that

$$f = P_1^+[\varphi] + P_1^-[\psi], \ g = -P_1^+[\psi] - P_1^-[\varphi].$$

The equations (12.14), (12.15) are not independent, because if $\varphi$ is any solution of (12.14) then the function $\psi := \varphi i i_3 = (P_3^+ - P_3^-)[\varphi]$ is a solution of (12.15) and vice versa.

The last step is to return from the function $\varphi$ satisfying (12.14) to the function $F$ satisfying (12.7), because (12.14) is the required biquaternionic equation which is equivalent to (12.7) and does not contain the operator $Z_{\mathrm{C}}$. Thus,

$$F = f + g^* = P_1^+[\varphi] + P_1^-[\varphi i i_3] - Z_{\mathrm{C}}[P_1^+[\varphi i i_3] + P_1^-[\varphi]] =$$

$$= (P_1^+ - P_1^+ Z_{\mathrm{C}} + P_1^-(P_3^+ - P_3^-) + P_1^-(P_3^+ - P_3^-)Z_{\mathrm{C}})[\varphi] =$$

$$= (P_1^+(I - Z_{\mathrm{C}}) + P_1^-(P_3^+ - P_3^-)(I + Z_{\mathrm{C}}))[\varphi];$$

we have taken advantage of the obvious commutation relations

$$P_k^\pm Z_{\mathrm{C}} = Z_{\mathrm{C}} P_k^\mp .$$

Conversely,

$$\varphi = (P_1^+ P_3^+ - Z_{\mathrm{C}} P_1^+ P_3^-)[F] = (P_1^+ P_3^+ - P_1^- P_3^+ Z_{\mathrm{C}})[F].$$

Let us write

$$\mathcal{U} := P_1^+ P_3^+ - P_1^- P_3^+ Z_{\mathrm{C}}, \ \ \mathcal{U}^{-1} := P_1^+(I - Z_{\mathrm{C}}) + P_1^-(P_3^+ - P_3^-)(I + Z_{\mathrm{C}}).$$

The final result is

$$N = \mathcal{U}^{-1} R \mathcal{U}, \tag{12.16}$$

where $R$ is the operator of equation (12.14):

$$R := P_1^+(i\partial_t + D) + P_1^-(-i\partial_t + D) - mM^{i_2}.$$

Of course, equality (12.16) can be verified also by a straightforward calculation.

From (12.8) and (12.7) we have

$$R = \mathcal{U} i \mathcal{A} \gamma_0 \mathbb{D} \mathcal{A}^{-1} \mathcal{U}^{-1} \tag{12.17}$$

or, using assertion 6 of Lemma 12.2,

$$R = \mathcal{U} \mathcal{A} \hat{i} \gamma_0 \mathbb{D} \mathcal{A}^{-1} \mathcal{U}^{-1}. \tag{12.18}$$

Thus the study of the Dirac operator $\mathbb{D}$ is equivalent to that of the quaternionic differential operator $R$. The homogeneous part of the latter is, in fact, a direct sum of two Füter–type operators.

**12.7** In what follows we limit ourselves to the time–harmonic situation, that is, we consider time–harmonic spinor fields $\Phi$:

$$\Phi(t, x) = q(x) e^{i\omega t}, \tag{12.19}$$

where $\omega \in \mathbb{R}; q := (q_0, q_1, q_2, q_3)^T : \mathbb{R}^3 \longrightarrow \mathbb{C}^4$. Then from (12.4) we obtain the equation for $q$,

$$\mathbf{D}_{\omega,m}[q] := (i\omega\gamma_0 - \sum_{k=1}^{3} \gamma_k \partial_k + im)[q] = 0, \tag{12.20}$$

which generally speaking must be satisfied in some domain $\tilde{\Omega} \subset \mathbb{R}^3$. We assume that $q \in C^1(\tilde{\Omega}) \cap C(\overline{\tilde{\Omega}})$, and $\partial\tilde{\Omega} =: \tilde{\Gamma}$ is a closed Liapunov surface in $\mathbb{R}^3$.

Now we look for the corresponding biquaternionic reformulation of equation (12.20). We must determine what happens with the function $\varphi := \mathcal{U}\mathcal{A}[\Phi]$ when the initial function $\Phi$ is time-harmonic.

**12.8 Lemma** *Let $\Phi$ be a time-harmonic function, i.e., of the form (12.19). Then*

$$\mathcal{A}[\Phi](t,x) = P_3^+\mathcal{A}[q](x)e^{i\omega t} + P_3^-\mathcal{A}[q](x)e^{-i\omega t}. \tag{12.21}$$

PROOF. A straightforward calculation for a function of the form (12.19) shows that

$$\mathcal{A}[\Phi](t,x) = \mathcal{A}[\cos(\omega t)q(x) + i\sin(\omega t)q(x)] =$$

$$= \cos(\omega t)\mathcal{A}[q](x) + \sin(\omega t)\mathcal{A}[iq](x).$$

Using assertion 5 of Lemma 12.2 we obtain

$$\mathcal{A}[\Phi](t,x) = \cos(\omega t)\mathcal{A}[q](x) - \sin(\omega t)\mathcal{A}[q](x)i_3.$$

Using the formulas

$$\cos(\omega t) = \frac{e^{i\omega t} + e^{-i\omega t}}{2}; \quad \sin(\omega t) = \frac{e^{i\omega t} - e^{-i\omega t}}{2i} \tag{12.22}$$

we have

$$\mathcal{A}[\Phi](t,x) = \frac{1}{2}(\mathcal{A}[q](x) + i\mathcal{A}[q](x)i_3)e^{i\omega t} + \frac{1}{2}(\mathcal{A}[q](x) - i\mathcal{A}[q](x)i_3)e^{-i\omega t} =$$

$$= P_3^+\mathcal{A}[q](x)e^{i\omega t} + P_3^-\mathcal{A}[q](x)e^{-i\omega t}.$$

□

**12.9** Now let us apply the transform $\mathcal{U}$ to the equality (12.20):

$$\mathcal{U}\mathcal{A}[\Phi] = e^{i\omega t}P_1^+P_3^+\mathcal{A}[q](x) - Z_C(e^{-i\omega t}P_1^+P_3^-\mathcal{A}[q](x)) = e^{i\omega t}\mathcal{U}\mathcal{A}[q](x).$$

This means that the time-harmonic function $\Phi$ is transformed to the time-harmonic function $F = e^{i\omega t}\rho(x)$. Moreover,

$$R[e^{i\omega t}\rho(x)] = -\omega(P_1^+ - P_1^-)[e^{i\omega t}\rho(x)] + D[e^{i\omega t}\rho(x)] - mM^{i_2}[e^{i\omega t}\rho(x)].$$

Therefore for the amplitude $\rho$ we have from (12.14) the equation

$$D_\alpha[\rho] = 0, \tag{12.23}$$

where $\alpha := -(i\omega i_1 + m i_2)$ is a complex quaternion.

In fact we have established the following direct correlation between $\mathbf{D}_{\omega,m}$ and $D_\alpha$:

$$\mathbf{D}_{\omega,m} = -\gamma_0 i \mathcal{A}^{-1}\mathcal{U}^{-1} D_\alpha \mathcal{U}\mathcal{A} \tag{12.24}$$

with $\alpha := -(i\omega i_1 + m i_2)$. Of course, this can be expressed in the following terms also:

**12.10 Theorem** *Equation (12.23) is equivalent to equation (12.20). If $q$ is a solution of (12.20), then $\rho := \mathcal{U}\mathcal{A}[q]$ satisfies (12.23) and, conversely, if $\rho \in \ker D_\alpha, \alpha := -(i\omega i_1 + m i_2)$ then $q := \mathcal{A}^{-1}\mathcal{U}^{-1}[\rho]$ is a solution of (12.20).*

**12.11** Observation. The fact that the study of the Dirac equation requires $\alpha$-holomorphic function theory with the quaternionic (not complex) $\alpha$ is quite unexpected. But it is even more remarkable, in our opinion, that the case of $\alpha$ being a zero divisor is not excluded, as a simple computation shows:

$$\alpha \in \mathfrak{S} \iff \omega^2 = m^2.$$

**12.12** Let us write down the map $\mathcal{U}\mathcal{A}$ in explicit form. We have

$$\mathcal{U}\mathcal{A}[\tilde{q}] = (P_1^+ P_3^+ - P_1^- P_3^+ Z_C)[(\mathrm{Re}q_0 + i\mathrm{Im}q_2)i_0 +$$

$$+(\mathrm{Im}q_1 - i\mathrm{Re}q_3)i_1 - (\mathrm{Re}q_1 + i\mathrm{Im}q_3)i_2 - (\mathrm{Im}q_0 - i\mathrm{Re}q_2)i_3].$$

Let us consider

$$P_3^+ \mathcal{A}[\tilde{q}] = \frac{1}{2} M^{(1+ii_3)} \mathcal{A}[\tilde{q}] =$$

$$= \frac{1}{2}[(\mathrm{Re}q_0 + i\mathrm{Im}q_2)i_0 + (\mathrm{Im}q_1 - i\mathrm{Re}q_3)i_1 - (\mathrm{Re}q_1 + i\mathrm{Im}q_3)i_2 - (\mathrm{Im}q_0 - i\mathrm{Re}q_2)i_3] +$$

$$+(\mathrm{Re}q_2 + i\mathrm{Im}q_0)i_0 + (\mathrm{Im}q_3 - i\mathrm{Re}q_1)i_1 - (\mathrm{Re}q_3 + i\mathrm{Im}q_1)i_2 - (\mathrm{Im}q_2 - i\mathrm{Re}q_0)i_3] =$$

$$= \frac{1}{2}[(q_0 + q_2)i_0 - i(q_1 + q_3)i_1 - (q_1 + q_3)i_2 + i(q_0 + q_2)i_3].$$

Then

$$P_1^+ P_3^+ \mathcal{A}[\tilde{q}] = \frac{1}{4} M^{(1+ii_1)}[(q_0 + q_2)i_0 - i(q_1 + q_3)i_1 - (q_1 + q_3)i_2 + i(q_0 + q_2)i_3] =$$

$$= \frac{1}{4}[(q_0 - q_1 + q_2 - q_3)i_0 + i(q_0 - q_1 + q_2 - q_3)i_1 -$$

$$-(q_0 + q_1 + q_2 + q_3)i_2 + i(q_0 + q_1 + q_2 + q_3)i_3].$$

By analogy,

$$P_3^+ Z_C \mathcal{A}[\tilde{q}] = \frac{1}{2}[(q_0 - q_2)i_0 - i(q_1 - q_3)i_1 - (q_1 - q_3)i_2 + i(q_0 - q_2)i_3]$$

and

$$P_1^- P_3^+ Z_C \mathcal{A}[\tilde{q}] = \frac{1}{4}[(q_0 + q_1 - q_2 - q_3)i_0 - i(q_0 + q_1 - q_2 - q_3)i_1 +$$

$$+(q_0 - q_1 - q_2 + q_3)i_2 + i(q_0 - q_1 - q_2 + q_3)i_3].$$

Then, finally, we have

$$\mathcal{U}\mathcal{A}[\tilde{q}] = \frac{1}{2}[-(q_1 - q_2)i_0 + i(q_0 - q_3)i_1 - (q_0 + q_3)i_2 + i(q_1 + q_2)i_3]. \tag{12.25}$$

In the next section we will need a simple consequence of the formula (12.25): how the map $\mathcal{U}\mathcal{A}$ acts on functions with only a pair of components differing from zero. We have, obviously,

$$\mathcal{U}\mathcal{A} : q = (q_0, 0, 0, q_3)^T \longmapsto \rho = \rho_1 i_1 + \rho_2 i_2, \tag{12.26}$$

where $\rho_1 = -\frac{1}{2i}(\tilde{q}_0 - \tilde{q}_3), \rho_2 = -\frac{1}{2}(\tilde{q}_0 + \tilde{q}_3)$ and

$$\mathcal{U}\mathcal{A} : q = (0, q_1, q_2, 0)^T \longmapsto \rho = \rho_0 i_0 + \rho_3 i_3, \tag{12.27}$$

where $\rho_0 = -\frac{1}{2}(\tilde{q}_1 - \tilde{q}_2), \rho_3 = -\frac{1}{2i}(\tilde{q}_1 + \tilde{q}_2)$.

Analogous formulas hold for the inverse map:

$$\mathcal{A}^{-1}\mathcal{U}^{-1} : \rho = \rho_1 i_1 + \rho_2 i_2 \longmapsto q = (-i\tilde{\rho}_1 - \tilde{\rho}_2, 0, 0, i\tilde{\rho}_1 - \tilde{\rho}_2) \tag{12.28}$$

and

$$\mathcal{A}^{-1}\mathcal{U}^{-1} : \rho = \rho_0 + \rho_3 i_3 \longmapsto q = (0, -\tilde{\rho}_0 - i\tilde{\rho}_3, \tilde{\rho}_0 - i\tilde{\rho}_3, 0). \tag{12.29}$$

Formulas (12.26)–(12.29) may be interpreted as follows: both $\mathcal{U}\mathcal{A}$ and $(\mathcal{U}\mathcal{A})^{-1}$ do not mix up the pairs of components which include the extreme ones or the middle ones.

**12.13** From (12.24) one can write down the map $\mathcal{UA}$ in the most simple and explicit form. Having identified the complex quaternion $\rho$ with the $\mathbb{C}^4$-vector $(\rho_0, \rho_1, \rho_2, \rho_3)^T$ one obtains the map $\mathcal{UA} : \tilde{q} \longmapsto \rho$ as

$$(\rho_0, \rho_1, \rho_2, \rho_3)^T = \mathcal{UA}[\tilde{q}] = \frac{1}{2}\begin{pmatrix} 0 & -1 & 1 & 0 \\ i & 0 & 0 & -i \\ -1 & 0 & 0 & -1 \\ 0 & i & i & 0 \end{pmatrix}\begin{pmatrix} q_0 \\ q_1 \\ q_2 \\ q_3 \end{pmatrix}. \tag{12.30}$$

By analogy, $\mathcal{A}^{-1}\mathcal{U}^{-1} : \tilde{\rho} \longmapsto q$ is equivalent to

$$(q_0, q_1, q_2, q_3)^T = \mathcal{A}^{-1}\mathcal{U}^{-1}[\tilde{\rho}] = \begin{pmatrix} 0 & -i & -1 & 0 \\ -1 & 0 & 0 & -i \\ 1 & 0 & 0 & -i \\ 0 & i & -1 & 0 \end{pmatrix}\begin{pmatrix} \rho_0 \\ \rho_1 \\ \rho_2 \\ \rho_3 \end{pmatrix}. \tag{12.31}$$

**12.14** Our constructions contain some formal asymmetry with respect to the space coordinates. This originates in the initial choice of the Dirac matrices in standard form. But of course it does not matter in what order we enumerate the space coordinates, and we can easily obtain the map $\mathcal{A}$ with the reflections generated by the axes $x_1$ and $x_2$. In this sense our model does conserve space symmetry.

# 13 Integral representations and boundary value problems for time-harmonic spinor fields

**13.1** The existence of such a simple and efficient map $\mathcal{UA}$, which establishes a close relationship between the functions from $\ker D_\alpha$ and time-harmonic massive spinor fields, allows one to extend the results of Chapter 1 onto the spinor fields. We are not going to transform all results obtained in that chapter to the case of spinor fields but to show those most interesting in our opinion. Once it is illustrated how this can be done, other facts may be obtained analogously.

**13.2** Let us introduce some necessary notation. Let

$$\mathbb{K}_{\omega,m} := \mathcal{A}^{-1}\mathcal{U}^{-1}K_\alpha \mathcal{U}\mathcal{A}, \tag{13.1}$$

$$S_{\omega,m} := \mathcal{A}^{-1}\mathcal{U}^{-1}S_\alpha \mathcal{U}\mathcal{A}, \quad \mathbb{P}_{\omega,m} := \frac{1}{2}(I + S_{\omega,m}), \tag{13.2}$$

$$\mathbb{T}_{\omega,m} := \mathcal{A}^{-1}\mathcal{U}^{-1}T_\alpha \mathcal{U}\mathcal{A}, \tag{13.3}$$

where $\alpha := -(i\omega i_1 + m i_2)$. The integrals in (13.2) exist in the sense of Cauchy's principal value.

An analogue of the Cauchy integral formula for time-harmonic massive spinor fields is as follows:

**13.3 Theorem** (Cauchy's integral formula for time-harmonic massive spinor fields) *Let* $q \in C(\bar{\Omega}) \cap \ker \mathbb{D}_{\omega,m}(\Omega)$. *Then*

$$q(x) = \mathbb{K}_{\omega,m}[q](x), x \in \Omega. \tag{13.4}$$

PROOF. From the hypothesis of the theorem it follows that

$$\mathcal{U}\mathcal{A}[q] =: \rho \in C(\bar{\tilde{\Omega}}) \cap \ker D_\alpha(\tilde{\Omega}), \qquad \alpha := -(i\omega i_1 + m i_2).$$

Then $\rho = K_\alpha[\rho]$ in $\tilde{\Omega}$ (Theorem 4.17, assertion 2). That is, $\mathcal{U}\mathcal{A}[q] = K_\alpha \mathcal{U}\mathcal{A}[q]$. Thus, $q = \mathcal{A}^{-1}\mathcal{U}^{-1}K_\alpha \mathcal{U}\mathcal{A}[q]$. □

Using Theorem 4.17 (assertion 4) we obtain the following

**13.4 Theorem** *If* $q \in C^1(\Omega) \cap C(\bar{\Omega})$, *then*

$$(\mathbb{D}_{\omega,m}\mathbb{T}_{\omega,m})[q] = -\gamma_0 \hat{i} q \text{ in } \Omega. \tag{13.5}$$

PROOF.

$$(\mathbf{D}_{\omega,m}\mathbf{T}_{\omega,m})[q] = -\gamma_0 \overset{?}{\imath} \mathcal{A}^{-1}\mathcal{U}^{-1} D_\alpha \mathcal{U}\mathcal{A}\mathbf{T}_{\omega,m}[q]$$
$$= -\gamma_0 \overset{?}{\imath} \mathcal{A}^{-1}\mathcal{U}^{-1} D_\alpha T_\alpha \mathcal{U}\mathcal{A}[q]. \tag{13.6}$$

Using Theorem 4.17 (assertion 4) we obtain from (13.6) the equality (13.5). □

By means of Theorem 5.9 on the holomorphic extension of a given $\mathbb{H}(\mathbb{C})$-valued function on a surface, we obtain the following

**13.5 Theorem** (On the spinor extension from the boundary) *Suppose $p : \Gamma \to \mathbb{C}^4$ satisfies the Hölder condition. In order for this function to be a boundary value of a solution $q$ to (12.19) in $\Omega$ it is necessary and sufficient that*

$$p = S_{\omega,m}[p] \text{ on } \Gamma. \tag{13.7}$$

*Moreover, if (13.7) is satisfied, then $q := \mathbf{K}_{\omega,m}[p]$ is the above-mentioned solution to (12.19).*

PROOF. This is completely analogous to the proofs of the two previous theorems. □

**13.6** Theorem 13.5 gives a complete description of the solvability picture of the Dirichlet problem for the equation (12.19): given a $\mathbb{C}^4$-valued function $p \in C^{0,\epsilon}(\Gamma)$, to find a function $q \in \ker\mathbf{D}_{\omega,m}(\Omega)$ such that

$$q|_\Gamma = p. \tag{13.8}$$

By Theorem 13.5 this problem is solvable if and only if condition (13.7) is satisfied. Moreover, if (13.7) is fulfilled, then the problem has the unique solution $q := \mathbf{K}_{\omega,m}[p]$.

**13.7** In Subsections 5.12–5.21 we constructed the Hilbert operator and obtained the Hilbert formulas for $\vec{\alpha}$-holomorphic functions. Now it is clear how to do the same for spinor fields.

Let $\Gamma := \mathbb{R}^2 = \{x \in \mathbb{R}^3 | x_3 = 0\}$, $\vec{n} = -i_3$. Introduce the notation

$$\mathbb{H}_{\omega,m} := -\mathcal{A}^{-1}\mathcal{U}^{-1}H_{\vec{\alpha}}M^{i_1}\mathcal{U}\mathcal{A} \tag{13.9}$$

where the operator $H_{\vec{\alpha}}$ is defined in Subsection 5.18. It is clear that its inverse is given by

$$\mathbb{H}_{\omega,m}^{-1} = \mathcal{A}^{-1}\mathcal{U}^{-1}M^{i_1}H_{\vec{\alpha}}^{-1}\mathcal{U}\mathcal{A}. \tag{13.10}$$

Taking into account the definitions of $\mathcal{U}$ and $\mathcal{A}$ we can conclude that formally both $\mathbb{H}_{\omega,m}$ and $\mathbb{H}_{\omega,m}^{-1}$ act on four-dimensional vectors of a specific form:

$$\mathbb{H}_{\omega,m} : (a_0, 0, 0, a_3) \longmapsto (0, b_1, b_2, 0),$$

$$\mathbb{H}_{\omega,m}^{-1} : (0, b_1, b_2, 0) \longmapsto (a_0, 0, 0, a_3),$$

which reflects the existence of a bicomplex number structure on $\mathbb{H}(\mathbb{C})$.

Now Theorem 5.19 together with formulas (12.25)–(12.28) imply that the Hilbert formulas for spinor fields hold.

**13.8 Theorem** (Hilbert formulas for spinor fields) *Given $\rho$ from $C^{0,\epsilon}(\mathbb{R}^2)$ or $L_p(\mathbb{R}^2)$, $\rho$ is a boundary value of a function $\varrho \in \ker\mathbf{D}_{\omega,m}(\mathbb{R}_3^+)$ if, and only if, pairs of components $(\rho_0, \rho_3)$ and $(\rho_1, \rho_2)$ are related by the formulas*

$$(0, \rho_1, \rho_2, 0)^T = \mathbb{H}_{\omega,m}[(\rho_0, 0, 0, \rho_3)^T], \tag{13.11}$$

$$(\rho_0, 0, 0, \rho_3)^T = \mathbb{H}_{\omega,m}^{-1}[(0, \rho_1, \rho_2, 0)^T]. \tag{13.12}$$

PROOF. This follows directly from the definitions of $\mathbb{H}_{\omega,m}^{\pm 1}$ and from Theorem 5.19.

□

**13.9 Corollary** (Reconstruction of the boundary value of a time-harmonic massive spinor field from two of its complex components) *Given one of the pairs of complex* *components* $(\rho_0, \rho_3)$ *or* $(\rho_1, \rho_2)$ *of the boundary value* $\rho(\in C^{0,\epsilon}(\mathbf{R}^2)$ *or* $\in L_p(\mathbf{R}^2))$ *of a function* $\varrho \in \ker\mathbf{D}_{\omega,m}(\mathbb{R}_3^+)$ . *Then the function* $\rho$ *is determined uniquely:*

$$\begin{aligned} \rho &= \mathbf{H}_{\omega,m}^{-1}[(0,\rho_1,\rho_2,0)^T] + (0,\rho_1,\rho_2,0)^T = \\ &= (\rho_0,0,0,\rho_3)^T + \mathbf{H}_{\omega,m}[(\rho_0,0,0,\rho_3)^T]. \end{aligned} \tag{13.13}$$

PROOF. This follows directly from Theorem 13.8. □

**13.10 Corollary** (Reconstruction of a time-harmonic massive spinor field by two complex components of its boundary value) *Given one of the pairs* $(\rho_0, \rho_3)$ *or* $(\rho_1, \rho_2)$ *(from the spaces* $C^{0,\epsilon}(\mathbf{R}^2)$ *or* $L_p(\mathbf{R}^2)$*) which are complex components of the boundary value of a function* $\varrho \in \ker\mathbf{D}_{\omega,m}(\mathbf{R}_3^+)$ . *Then the function* $\varrho$ *is determined uniquely:*

$$\begin{aligned} \varrho &= \mathbf{K}_{\omega,m}[\mathbf{H}_{\omega,m}^{-1}[(0,\rho_1,\rho_2,0)^T] + (0,\rho_1,\rho_2,0)^T] = \\ &= \mathbf{K}_{\omega,m}[(\rho_0,0,0,\rho_3)^T + \mathbf{H}_{\omega,m}[(\rho_0,0,0,\rho_3)^T]]. \end{aligned} \tag{13.14}$$

Thus, we are able to find a solution of the following

**13.11 Problem** Given two $\mathbb{C}$-valued functions $a, b$ on $\Gamma$, to find $\varrho \in \ker\mathbf{D}_{\omega,m}(\mathbf{R}_3^+)$ vanishing at infinity such that

$$\varrho_0|_\Gamma = a, \quad \varrho_3|_\Gamma = b. \tag{13.15}$$

We assume, as usual, that $\{a,b\} \subset C^{0,\epsilon}(\mathbf{R}^2)$ or $\{a,b\} \subset L_p(\mathbb{R}^2)$.

**13.12** Theorem 13.8 together with Corollary 13.10 shows that $\varrho$ is defined uniquely, and the solution is given as follows.

Firstly, we construct the function $\rho := (a,0,0,0,b)^T + \mathbf{H}_{\omega,m}[(a,0,0,b)^T]$. Then the solution of Problem 13.11 is given by

$$\varrho = \mathbf{K}_{\omega,m}[\rho].$$

**13.13** If we change the boundary condition (13.15) to

$$\varrho_1|_\Gamma = a, \quad \varrho_2|_\Gamma = b, \tag{13.16}$$

then $\varrho = \mathbf{K}_{\omega,m}[\rho]$, where

$$\rho = (0,a,b,0)^T + \mathbf{H}_{\omega,m}^{-1}[(0,a,b,0)^T].$$

# 14 Boundary integral criteria for the MIT bag model

**14.1** One of the most interesting problems in contemporary elementary particle physics is the problem of quark confinement. The commonly accepted hypothesis asserts that quarks are not found in a free state outside of a domain occupied by a hadron. The mechanism of their confinement to a hadronic bag is not yet understood, but there are some mathematical models describing this phenomenon more or less successfully.

One such mathematical option, which we will discuss in this section, is to complement the Dirac equation describing any particle of spin 1/2 (including quarks) with some additional conditions on the boundary of the domain occupied by a hadron. These conditions impose a ban on the particle flow through the surface of the confining region.

At present there exist many modifications of the original, so-called MIT (Massachusetts Institute of Technology) bag model; we will not survey them here. The MIT bag model proposed in [24, 23] remains the simplest and most useful for calculations, and taking it as an illustration we will show some applications of quaternionic analysis to this problem.

**14.2 Problem** (MIT bag model) To find a solution to the Dirac equation in $\tilde{\Omega}$,

$$\mathbf{D}_{\omega,m}[q] := (i\omega\gamma_0 - \sum_{k=1}^{3} \gamma_k \partial_k + im)[q] = 0, \tag{14.1}$$

satisfying the boundary condition

$$\sum_{k=1}^{3} \gamma_k \tilde{n}_k(x) q(x) = iq(x), \quad x \in \tilde{\Gamma}, \tag{14.2}$$

where $\tilde{n}_k$ are the components of the unit outward normal to the boundary $\tilde{\Gamma}$ at the point $x$. As earlier we assume that $q \in C^1(\tilde{\Omega}) \cap C(\bar{\tilde{\Omega}})$; $\tilde{\Gamma}$ is a closed Liapunov surface in $\mathbb{R}^3$.

**14.3** In Section 12.9 we established the precise relation between the operators $\mathbf{D}_{\omega,m}$ and $D_\alpha$ (formula (12.23)).

A function $q$ satisfies (14.1) if and only if its image $\rho := \mathcal{U}\mathcal{A}[q]$ under the map $\mathcal{U}\mathcal{A}$ satisfies the quaternionic equation

$$D_\alpha[\rho] = 0 \quad \text{in } \Omega, \tag{14.3}$$

where $\alpha := -(i\omega i_1 + m i_2)$.

We obtain the biquaternionic reformulation of the boundary condition (14.2) using the map $\mathcal{U}\mathcal{A}$. First of all we apply the mapping $\mathcal{A}$ to both sides of (14.2).

Now multiply (14.2) by $\gamma_0$ on the left-hand side. Then the equality

$$\mathcal{A}[(\sum_{k=1}^{3} \gamma_0 \gamma_k \tilde{n}_k q)|_{\tilde{\Gamma}}] = \mathcal{A}[(i\gamma_0 q)|_{\tilde{\Gamma}}]$$

is equivalent to

$$(\mathcal{A}[\sum_{k=1}^{3} \gamma_0 \gamma_k \tilde{n}_k q])|_{\Gamma} = (\mathcal{A}[i\gamma_0 q])|_{\Gamma}.$$

By Lemma 12.2, for the right-hand side we get

$$\mathcal{A}[i\gamma_0 q] = -(\mathcal{A}[q])^* i_3.$$

Analogously for the left-hand side we obtain, changing the normal vector $\bar{n}$ to $\bar{\Gamma}$ for $\vec{n}$, the normal vector to $\Gamma$,

$$(\mathcal{A}[\sum_{k=1}^{3} \gamma_0\gamma_k \bar{n}_k q])|_\Gamma = (i\vec{n}\mathcal{A}[q])|_\Gamma.$$

Thus, equality (14.2) becomes

$$i\vec{n}\mathcal{A}[q] + Z_C\mathcal{A}[q]i_3 = 0 \quad \text{on } \Gamma$$

or, in equivalent form,

$$\frac{1}{2}(I + \vec{n}iM^{i_3}Z_C)\mathcal{A}[q] = 0 \text{ on } \Gamma. \tag{14.4}$$

Let us write

$$Q^{\pm} := \frac{1}{2}(I \pm \vec{n}Z_C(P_3^+ - P_3^-)) = \frac{1}{2}(I \mp \vec{n}iM^{i_3}Z_C),$$

where $P_3^{\pm}$ are defined in Subsection 12.6.

Then (14.4) can be rewritten as

$$Q^-\mathcal{A}[q] = 0 \text{ on } \Gamma. \tag{14.5}$$

It is easy to verify that the operators $Q^+$ and $Q^-$ are mutually complementary projectors on a set of $\mathbb{H}(\mathbb{C})$-valued functions which is invariant with respect to them. Then the operators $S^{\pm} := \mathcal{U}Q^{\pm}\mathcal{U}^{-1}$ are also mutually complementary projectors.

It remains to obtain the explicit form of $S^{\pm}$.

From (12.16) it follows that

$$miZ_C M^{i_3} = \mathcal{U}^{-1} m M^{i_2}\mathcal{U};$$

that is, $\mathcal{U}(iM^{i_3}Z_C)\mathcal{U}^{-1} = M^{i_2}$. Then

$$S^{\pm} = \frac{1}{2}(I \mp \vec{n}\mathcal{U}(iM^{i_3}Z_{\mathbb{C}})\mathcal{U}^{-1}) = \frac{1}{2}(I \mp \vec{n}M^{i_2})$$

and, for $\rho := \mathcal{U}\mathcal{A}[q]$, equality (14.5) turns into the boundary condition

$$S^{-}[\rho] = 0 \quad \text{on} \quad \Gamma. \tag{14.6}$$

**14.4 Problem** (Biquaternionic reformulation of the MIT bag model)

To find an $\alpha-$holomorphic function, $\alpha = -(i\omega i_1 + mi_2)$, with the boundary condition (14.6).

**14.5** From the above reasoning, the MIT bag model and its biquaternionic reformulation are equivalent, and the equivalence is determined by the bijection $\mathcal{U}\mathcal{A}$.

If is clear that condition (14.6) can be rewritten in the form $\rho = S^{+}[\rho]$ on $\Gamma$. Theorem 5.7 now gives a simple criterion of solvability of Problem 14.4.

**14.6 Theorem** (Solvability of the MIT bag model problem) *Given* $\rho \in C^{0,\epsilon}(\Gamma; \mathbb{H}(\mathbb{C}))$, *Problem 14.4 is solvable if and only if*

$$\rho = P_{\alpha}[\rho] = S^{+}[\rho] \tag{14.7}$$

*holds on* $\Gamma$.

In this way the bag model (Problem 14.2) reduces to the system of two independent boundary equations (14.7).

This criterion can be slightly modified so that instead of two equalities in (14.7) we obtain only one.

**14.7 Theorem** *Given* $g \in C^{0,\epsilon}(\Gamma, \mathbb{H}(\mathbb{C}))$, *the function* $\rho := K_{\alpha}[g]$ *provides a solution to Problem 14.4 if and only if*

$$S^{-}P_{\alpha}[g] = 0 \tag{14.8}$$

*holds on* $\Gamma$.

PROOF. Let $\rho := K_\alpha[g]$ be a solution to Problem 14.4. By the quaternionic Plemelj–Sokhotski formula (5.2) we have:

$$S^-[\rho]|_\Gamma = S^- P_\alpha[g] = 0.$$

Suppose now, on the contrary, $S^- P_\alpha[g] = 0$ on $\Gamma$. Consider $\rho := K_\alpha[g]$. This satisfies (14.3) in $\Omega$, and on $\Gamma$ we have:

$$\rho = P_\alpha[g].$$

Therefore $S^-[\rho] = 0$ on $\Gamma$.

□

**14.8** REMARK Only the solutions $g$ of (14.8) which do not belong to $\ker P_\alpha$ are of interest because otherwise we obtain only the trivial solution for the bag model.

**14.9** To understand better what the condition (14.6) is, consider the simple example $\Gamma = \{x \in \mathbb{R}^3 | x_2 = 0\}$, $\vec{n} \equiv -\boldsymbol{i}_2$. Then

$$S^- = \frac{1}{2}(I - i_2 M^{i_2}).$$

Then condition (14.6) is equivalent to $\rho_0 = \rho_2 = 0$ on $\Gamma$.

In this situation evidently the results on the Hilbert formulas from Subsections 5.19–5.21 are applicable.

**14.10** In conclusion we note the following. Without appropriate integral representations for the massive spinor fields, the bag models could not have been studied quite fully. Up to now only some particular solutions of (14.1), (14.2) have been obtained for the case $\bar{\Omega}$ a ball (see, e.g., [124], [26, p. 413]) as well as for the case when $\bar{\Omega}$ is an infinite strip between two parallel hyperplanes [90, p. 58]. Biquaternionic analysis allows the development of methods for the treatment of boundary value problems relative to the operators $D_\alpha(\alpha \in \mathbb{H}(\mathbb{C}))$. Some of the simplest problems have been studied in

Section 6. The above considerations show that all results in this direction are applicable, in particular, to the analysis of quark models of the hadron structure. Thus, it is necessary to build up a systematic solvability and Fredholm theory in connection with analogues of the Hilbert and Riemann problems for the operator $D_\alpha$.

# Chapter 4

# Hypercomplex factorization, systems of non-linear partial differential equations generated by Füter–type operators

## 15 Notion of hypercomplex factorization and applications to mathematical physics

**15.1** In this section we are going to show that the Cauchy–Riemann operators $D_\alpha$ arise not only in the theories of the Maxwell and Dirac equations. There exists a quite general scheme involving a family of such operators (or, in other words, a matrix operator with entries consisting of these operators) which treats other problems of mathematical physics.

Denote by $\mathbb{H}^n(\mathbb{C}) := \mathbb{H}(\mathbb{C}) \times \cdots \times \mathbb{H}(\mathbb{C})$ the set of $n$-dimensional vectors with components from $\mathbb{H}(\mathbb{C})$. Let $\Omega$ be a bounded domain in $\mathbb{R}^3$ with the Liapunov boundary $\Gamma$, and let $\mathcal{F}(\Omega, \mathbb{H}^n(\mathbb{C}))$ denote some complex space (or $\mathbb{H}(\mathbb{C})$-module) of $\mathbb{H}^n(\mathbb{C})$-valued functions defined on $\Gamma$. For example, we shall often use an $\mathbb{H}(\mathbb{C})$-bimodule $\mathcal{F}(\Omega) := C^2(\Omega; \mathbb{H}^n(\mathbb{C})) \cap C^1(\bar{\Omega}; \mathbb{H}^n(\mathbb{C}))$.

On $C^1(\Omega; \mathbb{H}^n(\mathbb{C}))$ we define an operator matrix $\check{D}_\eta$ of the form

$$\check{D}_\eta = (\chi_{ij} D_{ij})_{i,j=1}^n \tag{15.1}$$

where $D_{ij} := D_{\eta_{ij}} := D + M^{\eta_{ij}}, \{\eta_{ij} | \forall i, j\} \subset \mathbb{H}(\mathbb{C}), \{\chi_{ij}\} \subset \{0, 1\}$.

**15.2 Definition** Let $\check{B}$ be an operator acting on $\mathcal{F}(\Omega; \mathbb{H}^n(\mathbb{C}))$. Then $\check{B}$ is said to admit a hypercomplex factorization if there exist two operator-matrices $\check{D}_\alpha$ and $\check{D}_\beta$ of the form (15.1), and a $\mathbb{C}$-linear operator $\check{A}$ such that

$$\check{B} = \check{D}_\alpha \check{A} \check{D}_\beta. \tag{15.2}$$

The fine point here is that initially $\check{B}$ does not necessarily have a "quaternionic nature", that is, does not necessarily require quaternions for its description. More precisely, $\check{B}$ may be a "real" or a "complex" operator in the sense that it maps, respectively, a $\mathbb{R}$-valued function onto a $\mathbb{R}$-valued one, or a $\mathbb{C}$-valued function onto a $\mathbb{C}$-valued one. Such an operator extends naturally on $\mathcal{F}$ in the component-wise manner (like, for instance, the Laplace operator), and hence we get an operator acting formally on $\mathbb{H}^n(\mathbb{C})$-valued functions extending a "non-exotic" real or "complex" operator.

In other words, $\check{B}$ may be a scalar or a vectorial differential operator defined via the "usual" differential operators div, rot, grad, etc. on $\mathbb{C}^{4n}$-valued functions, but the combination of those "usual" operators allows us to rewrite $\check{B}$ in the form (15.2), i.e., the study of $\check{B}$ reduces to that of the already known factors $\check{D}_\alpha$ and $\check{D}_\beta$, and of the arbitrary $\check{A}$.

It appears that quite a large number of operators from different areas of mathematical physics admit a hypercomplex factorization.

**15.3 Example** The Laplace operator in $\mathbb{R}^3$ gives us the simplest example: $n = 1$; $\alpha = \beta = 0$; $\check{A} = -I$.

**15.4 Example** The Helmholtz operator with both complex (thus being a complex operator in the above sense) and $\mathbb{H}(\mathbb{C})$-valued parameters admits a hypercomplex factorization: for $\Delta + M^{\lambda}$, $\lambda = \alpha^2$, we have $n = 1$, $\alpha \in \mathbb{H}(\mathbb{C})$, $\beta = -\alpha$, $\breve{A} = -I$.

**15.5 Example** Let $\mu$ and $\nu$ be the Lamé constants, then the Lamé operator

$$\breve{B} = L := \mu\Delta + (\nu + \mu)\text{graddiv} \tag{15.3}$$

acting on three-dimensional vector fields admits a hypercomplex factorization [46, p. 85]: $n = 1$; $\alpha = \beta = 0$,

$$\breve{A} = -\frac{1}{2}((\nu + 3\mu)I + (\nu + \mu)Z_{\mathbf{H}}).$$

**15.6 Example** The operator of steady oscillation has the form

$$\breve{B} := \mu\Delta + (\nu + \mu)\text{graddiv} + \rho\theta^2 = L + \rho\theta^2 \tag{15.4}$$

where $\rho$ is the density of the elastic medium; $\theta$ the oscillation frequency. Assume that $n = 1$, $\alpha = 0$, $\beta = \theta\frac{\sqrt{\rho}}{\sqrt{\mu}}$;

$$\breve{A} = -\frac{1}{2}((\nu + 3\mu)I + (\nu + \mu)Z_{\mathbf{H}}) + \theta\sqrt{\mu\rho}T_0,$$

then (15.2) holds. We shall verify this using Example 15.5. First,

$$D\breve{A}D_{\beta}[\vec{u}] = L[\vec{u}] + (-\frac{1}{2}(\nu + 3\mu)\beta D-$$

$$-\frac{1}{2}(\nu + \mu)\beta D Z_{\mathbf{H}} + \theta\sqrt{\mu\rho}DT_0D_{\beta})[\vec{u}].$$

Then using the properties of the right inverse $T_0$ we get

$$D\breve{A}D_{\beta}[\vec{u}] = L[\vec{u}] + (-\frac{1}{2}(\nu + 3\mu)\beta + \frac{1}{2}(\nu + \mu)\beta +$$

$$+\theta\sqrt{\mu\rho})D[\vec{u}] + \theta\sqrt{\mu\rho}\beta[\vec{u}] =$$

$$= L[\vec{u}] + (-\mu\beta + \theta\sqrt{\mu\rho})D[\vec{u}] + \rho\theta^2[\vec{u}].$$

Finally, the definition of $\beta$ yields

$$D\check{A}D_\beta = L + \rho\theta^2.$$

**15.7 Example** Up to now, all examples deal with the case $n = 1$. We now give an example for $n > 1$.

Consider the equations of statics in the theory of moment elasticity ([3, p. 189]):

$$(\mu + \gamma)\Delta[\vec{u}] + (\lambda + \mu - \gamma)\text{graddiv}[\vec{u}] + 2\gamma\text{rot}\vec{w} = 0, \tag{15.5}$$

$$(\nu + \theta)\Delta[\vec{w}] + (\epsilon + \nu - \theta)\text{graddiv}[\vec{w}] + 2\gamma\text{rot}\vec{u} - 4\gamma\vec{w} = 0, \tag{15.6}$$

where the coefficients $\lambda, \mu, \nu, \theta, \epsilon, \gamma$ characterizing the medium satisfy the conditions $\mu > 0$, $3\lambda + 2\mu > 0$; $\gamma > 0$, $\epsilon > 0$, $\theta > 0$, $3\epsilon + 2\nu > 0$, $\vec{u}$ is the displacement vector, and $\vec{w}$ the rotation vector.

Let us demonstrate that the operator defining the system (15.5)–(15.6) admits a hypercomplex factorization with $n = 2$. Denote by $\check{D}_\alpha$, $\check{D}_\beta$ and $\check{A}$ the following operators:

$$\check{D}_\alpha := \text{diag}\{D, D\},$$

$$\check{D}_\beta := \text{diag}\{D, D_{22}\}, \quad \beta_{22} := \frac{2i\sqrt{\gamma}}{\sqrt{\nu + 2\theta}},$$

$$\check{A} := \begin{pmatrix} A_{11} & A_{12} \\ A_{21} & A_{22} \end{pmatrix},$$

$$A_{11} := -\frac{1}{2}((\lambda + 3\mu + \gamma)I + (\lambda + \mu - \gamma)Z_H;$$

$$A_{12} := \gamma T_0(I - Z_{\mathbf{H}})U;$$

$$A_{21} := \gamma T_0(I - Z_{\mathbf{H}});$$

$$A_{22} := -\frac{1}{2}((\epsilon + 3\nu + 3\theta)I + (\epsilon + \nu - \theta)Z_{\mathbf{H}} - 4i\sqrt{\gamma(\nu + 2\theta)}T_0)$$

where $U$ acts on elements of the form $D_{22}[\vec{w}]$ by the rule $U[D_{22}[\vec{w}]] := D_{22}[\vec{w}] - \beta_{22}[\vec{w}]$.

Recalling Example 15.5 it is easy to see that

$$-\frac{1}{2}D((\lambda + 3\mu + \gamma)I + (\lambda + \mu - \gamma)Z_{\mathbf{H}})D[\vec{u}] = (\mu + \gamma)\Delta[\vec{u}] + (\lambda + \mu - \gamma)\mathbf{graddiv}\vec{u}.$$

In addition we have:

$$\gamma DT_0(I - Z_{\mathbf{H}})UD_{22}[\vec{w}] = \gamma(I - Z_{\mathbf{H}})D[\vec{w}].$$

Since $(I - Z_{\mathbf{H}})D[\vec{w}] = 2\mathrm{rot}\vec{w}$, (15.5) is obtained.

By analogy,

$$\gamma DT_0(I - Z_{\mathbf{H}})D[\vec{u}] = 2\gamma\mathrm{rot}\vec{u}.$$

Now let us consider

$$-\frac{1}{2}D((\epsilon + 3\nu + 3\theta)I + (\epsilon + \nu - \theta)Z_{\mathbf{H}} - 4i\sqrt{\gamma(\nu + 2\theta)}T_0)D_{22}[\vec{w}] =$$

$$= -\frac{1}{2}D((\epsilon + 3\nu + 3\theta)I + (\epsilon + \nu - \theta)Z_{\mathbf{H}}))D[\vec{w}] -$$

$$-\frac{1}{2}\beta_{22}D((\epsilon + 3\nu + 3\theta)I + (\epsilon + \nu - \theta))Z_{\mathbf{H}}[\vec{w}] +$$

$$+2i\sqrt{\gamma(\nu + 2\theta)}D[\vec{w}] + 2i\beta_{22}\sqrt{\gamma(\nu + 2\theta)}\vec{w} =:$$

$$=: J_1 + J_2 + J_3 + J_4.$$

$J_1$ (see Example 15.5) gives the first two summands of (15.6). Further,

$$((\epsilon + 3\nu + 3\theta)I + (\epsilon + \nu - \theta)Z_{\mathbf{H}})[\vec{w}] =$$

$$= (\epsilon + 3\nu + 3\theta)\vec{w} - (\epsilon + \nu - \theta)\vec{w} = 2(\nu + 2\theta)\vec{w}.$$

Therefore

$$J_2 = -\beta_{22}(\nu + 2\theta)D[\vec{w}] = -2i\sqrt{\gamma(\nu + 2\theta)}D[\vec{w}].$$

Hence

$$J_2 + J_3 = 0.$$

For the last summand $J_4$ we have

$$J_4 = 2i\beta_{22}\sqrt{\gamma(\nu + 2\theta)}\vec{w} = -4\gamma\vec{w},$$

which gives equation (15.6).

In the form $\check{D}_\alpha \check{A} \check{D}_\beta \; f = 0$ one may also represent the equations of steady oscillation in the theory of moment elasticity ([74], [3, p. 190]) and some other equations of mathematical physics.

**15.8** Let us stress once more that all the above examples contain well-known and important operators.

Now we provide some applications of Definition 15.2. We assume here that $\check{B}$ admits a hypercomplex factorization with the following properties:

- $\check{A}$ is continuously invertible,
- the operator $\check{D}_\beta$ has a right inverse $\check{T}_\beta$ such that there exist the limits

$$\lim_{\Omega \ni x \to \tau \in \Gamma} (I - \check{T}_\beta \check{D}_\beta)[f](x) =: \check{P}_\beta[f](\tau), \tag{15.7}$$

and

$$\lim_{\Omega \ni x \to \tau \in \Gamma} \check{T}_\beta \check{D}_\beta[f](x) =: \check{Q}_\beta[f](\tau), \tag{15.8}$$

and the operators $\check{P}_\beta$ and $\check{Q}_\beta$ generated by (15.7)–(15.8) satisfy the property

$$\check{P}_\beta + \check{Q}_\beta = I.$$

For instance, if $\check{D}_\beta$ is a diagonal matrix and $\chi_{ij} \neq 0$, $j = \overline{1,n}$ then a right inverse $\check{T}_\beta$ exists and possesses the required properties (this follows from Theorem 4.17). For Example 15.7, the operator $\check{T}_\beta = \text{diag}\{T_0, T_{22}\}$ with $\beta_{22} = 2i\dfrac{\sqrt{\gamma}}{\sqrt{\nu + 2\theta}}$.

Consider the boundary value problems

$$\check{B}[v](x) = 0, \quad x \in \Omega, \tag{15.9}$$

$$v|_\Gamma = h, \tag{15.10}$$

and

$$\check{D}_\beta[f](x) = 0, \quad x \in \Omega, \tag{15.11}$$

$$f|_\Gamma = \check{P}_\beta[h], \tag{15.12}$$

$$\check{D}_\alpha[g](x) = 0, \quad x \in \Omega, \tag{15.13}$$

$$(\check{T}_\beta \check{A}^{-1}[g])|_\Gamma = \check{Q}_\beta[h]. \tag{15.14}$$

Here we are looking for the functions $v, f, g$ from the relevant classes which satisfy the corresponding conditions. Thus we have: (15.9)–(15.10) is the Dirichlet problem for the operator $\check{B}$; (15.11)–(15.12) is also the Dirichlet problem but for the operator $\check{D}_\beta$ and with a special given function on the boundary; the type of the problem (15.13)–(15.14) depends on the properties of the operator $\check{A}$.

The following theorem, central in this section, establishes the precise relation between the solutions of the problem (15.9)–(15.10) and the pair of problems (15.11)–(15.12) and (15.13)–(15.14).

**15.9 Theorem** (Hypercomplex factorization and the Dirichlet problem) *Under the conditions of Subsection 15.8 there exists a one-to-one correspondence between the set of solutions to the problem (15.9)–(15.10) and the set of ordered pairs of solutions to* *the problems (15.11)–(15.12) and (15.13)–(15.14) which can be described as follows: if $\{f, g\} \subset \mathcal{F}(\Omega)$ are solutions to the problems (15.11)–(15.12) and (15.13)–(15.14) respectively, then the function*

$$v := f + \check{T}_\beta \check{A}^{-1}[g]$$

*is a solution to the problem (15.9)–(15.10), and vice versa: for any solution $v$ to the problem (15.9)–(15.10) the function*

$$f := (I - \check{T}_\beta \check{D}_\beta)[v]$$

*solves the problem (15.11)–(15.12) and the function*

$$g := \check{A}\check{D}_\beta[v]$$

*is a solution to the problem (15.13)–(15.14).*

PROOF. Let $f$ and $g$ be the solutions to the problems (15.11)–(15.12) and (15.13)–(15.14) respectively, and let $v := f + \check{T}_\beta \check{A}^{-1}[g]$. Then in $\Omega$

$$\check{D}_\alpha \check{A} \check{D}_\beta[v] = \check{D}_\alpha \check{A} \check{D}_\beta[f] + \check{D}_\alpha[g] = 0$$

by (15.11), (15.13), that is, $v$ satisfies (15.9). As for (15.10), we have

$$v|_\Gamma = (f + \check{T}_\beta \check{A}^{-1}[g])|_\Gamma = \check{P}_\beta[h] + \check{Q}_\beta[h] = h$$

which completes the first part of the proof.

Now let $v$ be a solution to (15.9)–(15.10). Consider $f := (I - \check{T}_\beta \check{D}_\beta)[v]$ and $g := \check{A}\check{D}_\beta[v]$. By the definition of $\check{T}_\beta$ we have

$$\check{D}_\beta[f] = \check{D}_\beta[v] - \check{D}_\beta \check{T}_\beta \check{D}_\beta[v] = 0$$

in $\Omega$, see (15.11). For the boundary $\Gamma$ we have

$$f|_\Gamma = (I - \check{T}_\beta \check{D}_\beta)[v]|_\Gamma = \check{P}_\beta[h],$$

see (15.12).

In conclusion we verify that $g$ satisfies (15.13)–(15.14). This follows from the chain of equalities

$$\check{D}_\alpha[g] = \check{D}_\alpha \check{A} \check{D}_\beta[v] = 0 \text{ in } \Omega,$$

and

$$\check{T}_\beta \check{A}^{-1}[g]|_\Gamma = \check{T}_\beta \check{D}_\beta[v]|_\Gamma = \check{Q}_\beta[h].$$

□

**15.10 Example** (See Example 15.3). The Dirichlet problem for the Laplace equation reduces to the pair of boundary value problems

$$\begin{cases} D[f](x) = 0, \quad x \in \Omega, \\ \\ f|_\Gamma = P_0[h], \end{cases}$$

$$\begin{cases} D[g](x) = 0, \qquad\qquad x \in \Omega, \\ \\ -(T_0[g])|_\Gamma = Q_0[h]. \end{cases} \tag{15.15}$$

The boundary condition (15.15) can be modified. By the quaternionic Cauchy integral formula, $g = K_0[g]$ in $\Omega$. Consequently, $T_0[g] = T_0 K_0[g]$. The operator $K_0$ allows the factorization (Theorem 4.18):

$$K_0 = DV_0.$$

Then $T_0[g] = T_0 D V_0[g]$. Using the quaternionic Borel–Pompeiu formula (Theorem 4.11) we obtain

$$T_0[g] = (I - K_0)V_0[g].$$

Then by the quaternionic Plemelj–Sokhotski formula (Theorem 5.3) we have:

$$T_0[g](x) \longrightarrow Q_0V_0[g](y),$$

when $\Omega \ni x \to y \in \Gamma$.

Thus, (15.15) turns into the equality

$$-Q_0V_0[g]|_\Gamma = Q_0[h] \text{ on } \Gamma, \tag{15.16}$$

which is already a boundary integral equation (cf [46, p. 75]).

Thus we have reduced the Dirichlet problem for a differential operator of the second order to a pair of Dirichlet–type problems for hyperholomorphic functions, i.e., for solutions of systems of first order possessing a well-developed function theory.

**15.11 Example** (See Example 15.4). For $\alpha = \alpha_0 \in \mathbb{R}$ in [45, 46] the Dirichlet problem for the Helmholtz equation reduced to the problems (15.11)–(15.14). Note that for $\alpha \in \mathbb{H}(\mathbb{C})$ not only for $\alpha \in \mathbb{C}$, the boundary condition (15.14) can be transformed (by analogy to Example 15.10) into a boundary integral equation (like the equality (15.16) and in contrast with (15.15) where the boundary value of the integral over a domain takes part).

The possibility of such a transformation is provided by Theorem 4.18 on the factorization of $K_\alpha$ for any $\alpha \in \mathbb{H}(\mathbb{C})$.

Let us examine, as an example, the case $\alpha \notin \mathfrak{S}$, $\vec{\alpha}^2 \neq 0$, $\alpha_0 \neq 0$. Then condition (15.14) has the form

$$-T_{-\alpha}[g]|_\Gamma = Q_{-\alpha}[h]. \tag{15.17}$$

In the domain $\Omega$ we have $g = K_\alpha[g]$. Consequently, by Theorem 4.18 we obtain in $\Omega$

$$-T_{-\alpha}[g] = (4\alpha_0\gamma)^{-1}T_{-\alpha}D_{-\alpha}(\Delta + M^{\bar{\alpha}^2})(V_{\xi_-} - V_{\xi_+})[g].$$

Now we use the quaternionic Borel–Pompeiu formula (Theorem 4.17) and the fact that the simple layer potentials $V_{\xi_-}[g]$ and $V_{\xi_+}[g]$ satisfy the corresponding Helmholtz equations:

$$-T_{-\alpha}[g] = (4\alpha_0\gamma)^{-1}(I - K_{-\alpha})((M^{\bar{\alpha}^2} - \xi_-^2)V_{\xi_-} - (M^{\bar{\alpha}^2} - \xi_+^2)V_{\xi_+})[g] =$$

$$= (I - K_{-\alpha})(P^+V_{\xi_+} + P^-V_{\xi_-})[g].$$

By the quaternionic Plemelj–Sokhotski formula (Theorem 5.3) we have

$$-T_{-\alpha}[g] \longrightarrow Q_{-\alpha}(P^+V_{\xi_+} + P^-V_{\xi_-})[g],$$

when $\Omega \ni x \to y \in \Gamma$. Hence equality (15.12) becomes

$$Q_{-\alpha}(P^+V_{\xi_+} + P^-V_{\xi_-})[g] = Q_{-\alpha}[h] \text{ on } \Gamma.$$

We can make here the same remark as at the end of Example 15.10.

**15.12** Remark A straightforward calculation using Theorem 15.9 shows that if $v$ is a solution of the Dirichlet problem for the Helmholtz equation, then we obtain

$$\Pi_\alpha[v] = \frac{1}{2\alpha}g; \quad \Pi_{-\alpha}[v] = f - \frac{1}{2\alpha}g - T_{-\alpha}[g],$$

where $f$ and $g$ are solutions of the corresponding problems (15.11)–(15.12) and (15.13)–(15.14).

**15.13** Now we show that the notion of hypercomplex factorization allows one to generalize the results of Section 2 (in particular, Theorem 2.13 on the decomposition of the space of metaharmonic functions) to a more general class of operators.

Let the operator $B$ admit a hypercomplex factorization (for the sake of simplicity we consider here the case $n = 1$): $B = D_\alpha A D_\beta$. We assume that $\{\alpha, \beta\} \subset \mathbb{C}$ though using the definitions and results of Section 2 one can examine the case of arbitrary $\{\alpha, \beta\} \subset \mathbb{H}(\mathbb{C})$ also. Let $\alpha \neq \beta$ and suppose the $\mathbb{C}$-linear operator $A$ enjoys the properties $AD = DA$ and $A^2 = A$. It is easy to give an example of such an $A$ not coinciding with the identity operator. For instance, consider the projection operators $P_k^\pm := \frac{1}{2} M^{(1 \pm i i_k)}$ which were introduced in Subsection 12.6.

Note that under our assumptions the operator $B$ can be rewritten as follows:

$$B = -A(\Delta + aD + bI), \tag{15.18}$$

where $a = -(\alpha + \beta)$, $b = -\alpha\beta$ and the operator $A$ commutes with the operator in brackets.

Introduce the following operators

$$\Pi^+ := \frac{1}{\alpha - \beta} A D_\alpha; \quad \Pi^- := I - \Pi^+ \tag{15.19}$$

which we consider on $\ker B$.

**15.14 Proposition** *The operators* $\Pi^\pm$ *are mutually complementary projectors on* $\ker B$.

**Proof.** It is enough to show that $(\Pi^+)^2 = \Pi^+$. For an arbitrary function $f \in \ker B$ we have

$$(\Pi^+)^2[f] = \frac{1}{(\alpha - \beta)^2} A D_\alpha A D_\alpha [f] = \frac{1}{(\alpha - \beta)^2} D_\alpha A D_\alpha [f] =$$

$$= \frac{1}{(\alpha - \beta)^2} (D_\alpha A D_\beta + (\alpha - \beta) D_\alpha A)[f] =$$

$$= \frac{1}{\alpha - \beta} A D_\alpha [f] = \Pi^+[f].$$

□

**15.15 Proposition** *The following equalities hold*

$$\text{im}\,\Pi^+ = A(\ker D_\beta), \tag{15.20}$$

$$\text{im}\,\Pi^- = \ker(AD_\alpha). \tag{15.21}$$

In other words the equality (15.17) signifies that a function $f$ belongs to the range of the projector $\Pi^+$ iff there exists a function $g$ from the null set of $D_\beta$ such that $f = A[g]$. Similarly, $f$ belongs to the range of $\Pi^-$ iff it is a solution of the equation $AD_\alpha f = 0$.

PROOF. Let us prove first equality (15.20). Assume that

$$f \in A(\ker D_\beta), i.e., \exists g \in \ker D_\beta : f = A[g].$$

Then

$$\Pi^-[f] = A[g] - \frac{1}{\alpha-\beta} AD_\alpha A[g] =$$
$$= A[g] - \frac{1}{\alpha-\beta}(AD_\beta[g] + (\alpha-\beta)A[g]) = 0.$$

Therefore $A(\ker D_\beta) \subset \text{im}\Pi^+$.

For $f \in \text{im}\Pi^+$ we have

$$f = \frac{1}{\alpha-\beta} AD_\alpha[\varphi],$$

where $\varphi \in \ker B$. Then choosing

$$g := \frac{1}{\alpha-\beta} AD_\alpha[\varphi]$$

we see that $g \in \ker D_\beta$ and $f = A[g]$. Hence, $f \in A(\ker D_\beta)$. Equality (15.20) is proved.

To verify (15.21) it is enough to note that $\ker(AD_\alpha) = \ker\Pi^+$ and use Proposition 15.14. □

The two previous propositions imply the following result.

**15.16 Theorem** (A decomposition of the kernels of a class of operators admitting a hypercomplex factorization) *For an operator $B$ of the form (15.18) with the operator $A$ possessing the properties $AD = DA$, $A^2 = A$ the following representation of the kernel holds*

$$\ker B = \ker(AD_\alpha) \oplus A(\ker D_\beta). \tag{15.22}$$

**15.17** It is clear that equality (15.19) facilitates the finding of integral representations for functions from $\ker B$. In particular, when $A = I$ one can apply the reasoning of Theorem 4.21 without change.

# 16 Self-duality equation and other systems of non-linear partial differential equations generated by the Füter–type operators

**16.1** We have seen that various versions of quaternionic analysis work quite successfully for the treatment of different linear problems of mathematical physics. Here we show that the quaternionic analysis tool is of use for some non-linear problems also. More precisely, we give a description of a class of instantons associated with the Jackiw-Nohl-Rebbi-'t Hooft ansatz. It is not possible to describe all these terms here (because this would lead us too far from the main line of the book) but we refer the reader to the fundamental work [47] where all necessary details can be found.

In particular, in [47, p. 99] one can find the equation

$$\partial_t g + D[g] + g \cdot \bar{g} = 0 \tag{16.1}$$

where $D$ is our old acquaintance, the Moisil–Theodoresco operator, the unknown function $g$ is $\mathbb{H}(\mathbb{C})$-valued and defined in a domain $\Xi = I \times \Omega$, $I$ is some interval on $\mathbb{R}$, and $\Omega$ a domain in $\mathbb{R}^3$. Equation (16.1) is written in [47, p. 99] in just such a quaternionic form, with the detailed explanation of how (16.1) is obtained from the general self-duality equation taking the Jackiw-Nohl-Rebbi-'t Hooft ansatz for the gauge potential.

A solution to (16.1) is called an instanton associated with the Jackiw-Nohl-Rebbi-'t Hooft ansatz. Our aim is to prove that the set of solutions of a Füter–type equation generates a class of such instantons. This aim requires some preliminary work.

**16.2** The ideas we shall use are related to symmetry properties of differential equations. In a series of works ([33, 117, 80, 122] and others) there was established a close connection between, on the one hand, the so-called differential substitutions of the type of Cole–Hopf and Miura transformations, and, on the other hand, classic symmetries of differential equations. The remarkable book [95] is highly recommended to the reader who wants to go more profoundly into the subject of Lie groups and their applications to differential equations.

We need the following definition from [95, p.96].

**16.3 Definition** Let $L$ be a system of differential equations. Let $G$ be a local group of transformations which acts on an open subset $\Lambda$ of the space of independent and dependent variables of $L$ and which has the property that whenever $u = f(x)$ is a solution to $L$ and whenever $g \in G$ is such that $g \circ f$ is defined, then $g \circ f$ is a solution to $L$ also. Then $G$ is called a symmetry group of $L$.

**16.4** The following illustrative example has been taken from [122]. The heat equation in one space-variable is of the form

$$u_t = u_{\zeta\zeta} \tag{16.2}$$

where $\zeta \in \mathbb{R}$, $u_\zeta := \frac{\partial u}{\partial \zeta}$. Let $G$ denote the group of all dilatations $u \mapsto \lambda u$, $\lambda \in \mathbb{C}$. $G$ is a symmetry group of (16.2) because (16.2) is linear.

As can be easily established, any differential invariant of $G$ is a function of variables

$$t, \zeta, \frac{u_\zeta}{u}, \partial_\zeta(\frac{u_\zeta}{u}), \ldots, \partial_\zeta^j(\frac{u_\zeta}{u}), \ldots \ . \tag{16.3}$$

Recall that any equation of the form

$$u_t = F(t, \zeta, u, \partial_\zeta u, \ldots, \partial_\zeta^n u), n \geq 2, \tag{16.4}$$

is sometimes called *the evolution equation*. It is important for our purposes that after having made the substitution

$$\bar{t} = t, \quad \bar{\zeta} = \zeta, \quad \bar{u} = \frac{u_\zeta}{u} \tag{16.5}$$

in (16.2) we arrive at an equation of the form (16.4), i.e., to the evolution equation. This is a corollary of the fact that the group $G$ preserves the equation and the operator $\partial_t$ sends invariants to invariants.

(16.5) is the well-known Cole–Hopf substitution, and the corresponding evolution equation is nothing but the Burgers equation

$$\bar{u}_{\bar{t}} = \bar{u}_{\bar{\zeta}\bar{\zeta}} + 2\bar{u}\bar{u}_{\bar{\zeta}}. \tag{16.6}$$

**16.5** The above example describes, in fact, the main idea of the so-called "procedure of factorization" [122] which is defined in terms of invariants of classic symmetries of the initial equation. It is quite unfortunate that we must use the word "factorization" for something which has no connection whatever with a decomposition of something into the product of several factors. The terms "hypercomplex factorization" and "procedure of factorization" have no relation to one another.

There is no complete theory of the procedure of factorization. We give here only a simplified description, in accordance with the definition in [128] and [122].

Assume that we deal with a simple enough (or well-studied) equation $E$. Then one can construct all equations arising via substitutions of all symmetry group invariants of $E$. This procedure has been named "the procedure of factorization".

If the class of equations obtained via this procedure includes some "interesting" new objects, one can find their solutions starting from the solutions of the initial equation $E$.

It appears that the most transparent situation exists for "the procedure of factorization" for the simplest class of p.d.e., namely, the evolution equations involving one space variable and one time variable. [122] contains a systematic study of it.

The picture is far more sophisticated in the case of several variables. It is very hard to single out a "lucky" situation in order for the result of "the procedure of factorization" to have the same number of independent variables as the initial equation.

Now we show that such situations do occur and they are worth examination.

**16.6** In Subsection 3.24 we introduced the Füter–type equation

$$(\partial_t - aD)[u] = 0 \tag{16.7}$$

where $u \in C^1(\Xi; \mathbb{H}(\mathbb{C}))$, $\Xi = I \times \Omega$ for some interval $I \subset \mathbb{R}$ and some domain $\Omega \subset \mathbb{R}^3$; $a$ is a complex constant.

Note that (16.7) can be regarded as an evolution equation with several variables (see (16.4)).

We shall demonstrate that "the procedure of factorization" for (16.7) with respect to some subgroups of the symmetry group leads to an equation which is more general than (16.1). Thus we shall be able to obtain a new series of solutions to (16.1) starting from that of (16.7).

We shall use the subgroup of dilatations $u \mapsto \lambda u$, $\lambda \in \mathbb{C}$, which is obviously the symmetry subgroup of (16.7).

**16.7** Let $\psi \in C^2(\Omega, \mathbb{C})$, $\vec{h} := \operatorname{grad}\psi : \Omega \to \mathbb{C}^3$,

$$\Delta[\psi] + \langle \operatorname{grad}\psi, \operatorname{grad}\psi \rangle = 0. \tag{16.8}$$

As can be easily seen, if $\phi_0$ is an arbitrary harmonic function and

$$\text{grad}\psi := \frac{\text{grad}\phi_0}{\phi_0}, \tag{16.9}$$

then $\psi$ is a solution to (16.8). Let $b$ be a complex constant, $\mu$ be a complex quaternion, $\mu \in \mathbb{H}(\mathbb{C})\setminus(\mathfrak{S} \cup \{0\})$; and let $(\mu, v) := \sum_{k=0}^{3} \mu_k v_k$ for two complex quaternions $\mu, v$.

We are ready to consider the equation

$$\partial_t v = aD[v] + a\vec{h}v - \frac{a}{b}(\mu, v)v \tag{16.10}$$

for the unknown function $v \in C^1(\Omega; \mathbb{H}(\mathbb{C}))$. Note that assuming $a = 1$, and formally changing the functional $(\mu, v)$ for the unit one, we get on the right–hand side of (16.10) the operator

$$D(p_0, \nu) := D - (\frac{\text{grad}p_0}{p_0})I + \nu I$$

with the non-vanishing scalar function $p_0 \in C^1(\Omega; \mathbb{C})$ and the constant $\nu \in \mathbb{C}$.

The latter operator was introduced and studied in [120]. It admits the representation

$$D(p_0, \nu) = p_0 D_\nu p_0^{-1}. \tag{16.11}$$

This allowed us to construct in [120] the function theory corresponding to $D(p_0, \nu)$ making use of the function theory for $D_\nu$.

To study equation (16.10) we need the following fact.

**16.8 Lemma** *Given $\mu \in \mathbb{H}(\mathbb{C})$ and a function $\vec{h}$ as in Subsection 16.7; there exists a function $u \in C^2(\Xi; \mathbb{H}(\mathbb{C})) \cap \ker(\partial_t - aD)$ such that*

$$\vec{h} = \frac{\text{grad}_x(\mu, u)}{(\mu, u)}. \tag{16.12}$$

PROOF. Let $\psi$ be the function from the definition of $\vec{h}$. Then for any $c \in \mathbb{C}$, $\phi_0 := exp(\psi + c)$ satisfies (16.9). Besides, the condition (16.8) ensures that $\Delta[\phi_0] = 0$.

It remains to show that for any harmonic function $\phi_0$ there exists a function $u \in \ker(\partial_t - aD)$ such that $\phi_0 = (\mu, u)$. As $(\mu, u) = \text{Sc}(u\bar{\mu})$, we can conclude that there exist three complex-valued functions $\phi_1, \phi_2, \phi_3$ of four variables such that for $\phi := \phi_0 + \sum_{k=1}^{3} \phi_k i_k$ we have

$$u\bar{\mu} = \phi. \tag{16.13}$$

By the assumptions, $\mu$ is invertible and thus the condition $u \in \ker(\partial_t - aD)$ is equivalent to $\phi \in \ker(\partial_t - aD)$.

It remains only to settle the problem of existence, for a harmonic function $\phi_0$, of $\vec{\phi} := \phi_1 i_1 + \phi_2 i_2 + \phi_3 i_3$ with the property $\phi := \phi_0 + \vec{\phi} \in \ker(\partial_t - aD)$.

Reasoning as in [133] it is easy to see the following: any function $\phi \in \ker(\partial_t - aD)$ has components of the form

$$\begin{aligned}
\phi_0 &= \partial_2\theta + \partial_3\omega; \quad & \phi_1 &= -\partial_2\omega + \partial_3\theta; \\
\phi_2 &= a\partial_t\theta + \partial_1\omega; \quad & \phi_3 &= -\partial_1\theta + a\partial_t\omega,
\end{aligned}$$

where $\theta, \omega$ are two arbitrary functions in $\ker(\partial_t^2 + a^2\Delta)$. Choosing now an arbitrary function from $\ker(\partial_t^2 + a^2\Delta)$ as $\omega$ and then $\theta$ as a solution of the equation $\partial_2\theta = \phi_0 - \partial_3\omega$, we have the existence of $\vec{\phi}$. □

**16.9 Remark** In particular, it follows from the proof that there exist such functions $u$ which may depend on the variable $t$ but nevertheless the quotient

$$\frac{\text{grad}_x(\mu, u)}{(\mu, u)}$$

is constant with respect to $t$, and thus $\vec{h}$ depends on $x$ only. Moreover, we have proved, in fact, that there exists a family of functions $u$ satisfying (16.12) which is parametrizable by all scalar solutions of $\partial_t^2 w + a^2\Delta w = 0$.

**16.10** Again let $u$ be a solution of (16.7), $b$ an arbitrary complex number. Consider the following invariant of the group of dilatations:

$$v = \frac{bDu}{(\mu, u)}. \tag{16.14}$$

We assume that $u$ satisfies the condition $(\mu, u) \neq 0$. In the following theorem we show that (16.14) gives a family of solutions for the equation (16.10).

**16.11 Theorem** *Let the function $u \in C^2(\Xi) \cap \ker(\partial_t - aD)$ satisfy the condition (16.12). Then the function (16.14) is a solution of the equation (16.10).*

PROOF. Calculating

$$D[v] = -\frac{\mathbf{grad}(\mu, u)}{(\mu, u)} v + \frac{bD^2[u]}{(\mu, u)}$$

and

$$\partial_t v = -\frac{a}{b}(\mu, v)v + ab\frac{D^2[u]}{(\mu, u)}$$

and taking into account (16.12), we find that $v$ satisfies (16.10). □

Recalling about the non–commutativity of quaternions let us introduce another invariant of the group of dilatations:

$$w = \frac{bD_r[u]}{(\mu, u)},$$

where as above $D_r[u] := \sum_{k=1}^{3} \partial_k u i_k$.

If $u$ here belongs to $\ker(\partial_t - aD_r)$ then $w$ solves the equation

$$\partial_t w = aD_r[w] + a\vec{h}w - \frac{a}{b}(\mu, w)w.$$

Restricting ourselves to the special case $(\mu, u) := u_0$, that is, $\mu = i_0$, we obtain the following

**16.12 Theorem** *Let $u \in C^2(\Xi) \cap \ker(\partial_t - aD)$ and $u_0 \neq 0$. Then the function*

$$g := v + w = \frac{b}{u_0}(D + D_r)[u]$$

*belonging to $C^1(\Xi)$ is a solution of the equation*

$$\partial_t g = aD[g] - \frac{a}{2b} g\bar{g}. \tag{16.15}$$

PROOF. Let us consider

$$D[v] = b(D[u_0^{-1}]D[u] + u_0^{-1}D^2[u]).$$

Calculating

$$D[u_0^{-1}] = -\frac{1}{2}u_0^{-2}(D[u] - \overline{D_r[u]}) = -\frac{1}{2bu_0}(v - \bar{w})$$

we obtain

$$D[v] = -\frac{1}{2b}(v - \bar{w})v + \frac{b}{u_0}D^2[u].$$

Then we have

$$\partial_t v = b\{-\frac{a\mathrm{Sc}(D[u])}{bu_0}v + \frac{aD^2[u]}{u_0}\} = -\frac{a}{b}v_0 v + \frac{ab}{u_0}D^2[u];$$

i.e.,

$$\partial_t v = aD[v] - \frac{a}{2b}(\bar{v} + \bar{w})v. \tag{16.16}$$

By analogy, for the function $w$ we obtain

$$\partial_t w = aD[w] - \frac{a}{2b}(\bar{v} + \bar{w})w. \tag{16.17}$$

It remains to add (16.16) to (16.17), and we obtain (16.15).

□

**16.13 Corollary** (Instantons for the Jackiw-Nohl-Rebbi-'t Hooft ansatz as the gauge potential, and the Füter equation). *For any* $u \in C^2(\Xi; \mathbb{H}(\mathbb{C})) \cap \ker(\partial_t + D)$, $u_0 \neq 0$, *the function*

$$g := -\frac{1}{2u_0}(D[u] + D_r[u])$$

*is a solution of (16.1).*

PROOF. Take $a = -1$ and $b = -\frac{1}{2}$ in Theorem 16.12. □

**16.14** Applying "the procedure of factorization" to the Füter equation we obtain a new class of instantons. This means, in particular, that many properties of instantons can be derived from those of solutions of the Füter equation.

# Appendices

## 1 Real quaternions

**A1.1** Above, in Section 1, we gave a brief account of properties of quaternions which are necessary for our analytic purposes. But the set of quaternions is much richer structurally, and we believe that this richness will lead to further analytic consequences. That is why we present in Appendices A1–A3 an ample survey of quaternions independently of what has been written in Section 1.

**A1.2** Consider the real four-dimensional vector space $\mathbb{R}^4$ with a canonical basis

$$i_0 = (1,0,0,0), \quad i_1 = (0,1,0,0), \quad i_2 = (0,0,1,0), \quad i_3 = (0,0,0,1).$$

If $a, b \in \mathbb{R}^4$, $a = \sum_{k=0}^{3} a_k i_k$, $b = \sum_{k=0}^{3} b_k i_k$, with $\{a_k, b_k\}$ real numbers, then of course

$$a + b := \sum_{k=0}^{3} (a_k + b_k) i_k$$

and for any $\lambda \in \mathbb{R}$

$$\lambda a := \sum_{k=0}^{3} (\lambda a_k) i_k.$$

It is known from R. Hamilton that a "good" multiplication in $\mathbb{R}^4$ can be introduced as follows. It is enough to define it for the base elements which can be conveniently described by the following table (sometimes called the Cayley table):

|  | $i_0$ | $i_1$ | $i_2$ | $i_3$ |
|---|---|---|---|---|
| $i_0$ | $i_0$ | $i_1$ | $i_2$ | $i_3$ |
| $i_1$ | $i_1$ | $-i_0$ | $i_3$ | $-i_2$ |
| $i_2$ | $i_2$ | $-i_3$ | $-i_0$ | $i_1$ |
| $i_3$ | $i_3$ | $i_2$ | $-i_1$ | $-i_0$ |

(A1.1)

i.e., $i_0$ acts as identity, and

$$\begin{aligned} i_1 i_2 &= -i_2 i_1 = i_3, \\ i_2 i_3 &= -i_3 i_2 = i_1, \\ i_3 i_1 &= -i_1 i_3 = i_2, \\ i_k^2 &= -i_0, \quad k \in \mathbb{N}_3. \end{aligned} \tag{A1.2}$$

We mention here that in many books and articles there is a tradition (which can be traced to R. Hamilton himself) of denoting the imaginary units $i_k$ by

$$i_1 = i, \quad i_2 = k, \quad i_3 = j,$$

thus writing an arbitrary real quaternion as

$$q = a_0 \cdot 1 + a_1 i + a_2 j + a_3 k.$$

This multiplication extends onto the whole of $\mathbb{R}^4$ by linearity:

$$\begin{aligned} a \cdot b = (a_0 i_0 + a_1 i_1 + a_2 i_2 + a_3 i_3) \cdot (b_0 i_0 + b_1 i_1 + b_2 i_2 + b_3 i_3) = \\ = (a_0 b_0 - a_1 b_1 - a_2 b_2 - a_3 b_3) i_0 + \\ +(a_0 b_1 + a_1 b_0 + a_2 b_3 - a_3 b_2) i_1 + \\ +(a_0 b_2 + a_2 b_0 + a_3 b_1 - a_1 b_3) i_2 + \\ +(a_0 b_3 + a_3 b_0 + a_1 b_2 - a_2 b_1) i_3. \end{aligned} \tag{A1.3}$$

Then $\mathbb{R}^4$ becomes a real, associative, non-commutative algebra, called the algebra $\mathbb{H}$ of real quaternions.

Note that if $G\mathbb{H} := \mathbb{H}\backslash\{0\}$ then $(G\mathbb{H}, \cdot)$ is a non-abelian multiplicative group; $(\mathbb{H}, +, \cdot)$ is a skew field. More reasons for this can be found below.

The marvelous peculiarity of the rules (A1.2) is that, according to the famous Frobenius theorem, this is the only mode of introducing multiplication in $\mathbb{R}^4$, in such a way that we arrive at an algebra without zero divisors. Moreover, the same theorem says that there are no algebras without zero divisors in $\mathbb{R}^3$.

**A1.3** The set of complex numbers, an obvious predecessor of quaternions, has a very important operation, an automorphism termed "complex conjugation". In particular, just complex conjugation provides an opportunity to present a positive quadratic form $x^2 + y^2 = |z|^2$ as a product of two linear forms: $x^2 + y^2 = (x + iy)(x - iy) = z\bar{z}$.

The skew field $\mathbb{H}$ has an exact analogue of that operation. For $a \in \mathbb{H}$ let

$$\bar{a} := a_0 - \sum_{k=1}^{3} a_k i_3.$$

$\bar{a}$ will be called conjugate, or more precisely, quaternionic conjugate to $a$. Direct verification shows that $\forall a, b \in \mathbb{H}$

a) $\bar{\bar{a}} = a$,

b) $\overline{ab} = \bar{b} \cdot \bar{a}$,

c) $\overline{a + b} = \bar{a} + \bar{b}$,

d) $|\bar{a}| = |a|$.

Thus the mapping

$$Z_{\mathbb{H}} : a \in \mathbb{H} \longmapsto \bar{a} \in \mathbb{H} \tag{A1.4}$$

is an antiautomorphism of $\mathbb{H}$. If $|a|$ denotes the usual Euclidean norm of a four-dimensional vector $a \in \mathbb{R}^4$, $|a| := (\sum_{k=0}^{3} a_k^2)^{1/2}$, then clearly

$$a \cdot \bar{a} = \bar{a} \cdot a = |a|^2, \tag{A1.5}$$

$$(a_0 + a_1 i_1 + a_2 i_2 + a_3 i_3)(a_0 - a_1 i_1 - a_2 i_2 - a_3 i_3) = a_0^2 + a_1^2 + a_2^2 + a_3^2.$$

Again, as in the complex numbers case, one gets a positive quadratic form (now of four variables) as a product of two linear forms.

Note that $(\mathbb{H}, Z_{\mathbb{H}})$ is an algebra with involution.

For $a \in \mathbb{H}$ the number $|a|$ is usually called a modulus of the quaternion $a$. (A1.5) allows us to prove easily that

$$|a \cdot b| = |b \cdot a| = |a| \cdot |b|.$$

In fact,

$$|ab|^2 = ab \cdot \overline{ab} = ab \cdot \bar{b}\bar{a} = a \cdot |b|^2 \bar{a} = |b|^2 a \cdot \bar{a} = |b|^2 \cdot |a|^2.$$

The antiautomorphism $Z_{\mathbb{H}}$ can be represented as a product

$$Z_{\mathbb{H}} = Z_1 Z_2 Z_3 = Z_1 Z_3 Z_2 = \ldots = Z_3 Z_2 Z_1$$

of operations $Z_1, Z_2, Z_3$ : $Z_k$ changes the imaginary unit $i_k$ to $-i_k$, $Z_1[a] := a_0 - a_1 i_1 + a_2 i_2 + a_3 i_3$, etc. Sometimes we will call $Z_k$ "$k$th partial conjugation". They are obviously linear involutions, $Z_k(a+b) = Z_k(a) + Z_k(b)$, $Z_k[Z_k[a]] = a$, but $Z_k(a \cdot b) = Z_k(a) \cdot Z_k(b)$. Thus each $Z_k$ is an automorphism (not antiautomorphism) of $\mathbb{H}$.

**A1.4** Let $a \in G\mathbb{H}$, i.e., $a \neq 0$. Then

$$a^{-1} := \frac{\bar{a}}{|a|^2} \tag{A1.6}$$

is an inverse of $a$. As always, $(ab)^{-1} = b^{-1} \cdot a^{-1}$.

**A1.5** For $a = \sum_{k=0}^{3} a_k i_k$ denote $a_0 =: \mathrm{Sc}(a)$, $\sum_{k=1}^{3} a_k i_k =: \vec{a} =: \mathrm{Vec}(a)$ and call them the scalar and the vectorial part of the quaternion $a$, respectively. Sometimes they are also called the real and the imaginary part of $a$, but because of the need to employ complex numbers and complex quaternions we prefer these terms.

The set $\tilde{\mathbb{R}} := \{a | a \in \mathbb{H}, \mathrm{Vec}(a) = 0\}$ can be naturally identified with $\mathbb{R}$, the real numbers. Moreover the restrictions of all operations in $\mathbb{H}$ onto $\tilde{\mathbb{R}}$ coincide with the corresponding operations in $\mathbb{R}$. Thus there is no need to distinguish $\tilde{\mathbb{R}}$ and $\mathbb{R}$, and it is common to consider $\mathbb{R}$ as a (commutative) subfield of $\mathbb{H}$.

Of course, the set $\tilde{\mathbb{R}}^3 := \{a | a \in \mathbb{H}, \mathrm{Sc}(a) = 0\}$ is a Euclidean subspace of $\mathbb{H}$, but not a subalgebra. It is also identified with $\mathbb{R}^3$, and under these agreements

$$\mathbb{H} = \mathbb{R} \oplus \mathbb{R}^3 = \mathbb{R} \oplus i_1\mathbb{R} \oplus i_2\mathbb{R} \oplus i_3\mathbb{R}, \tag{A1.7}$$

$\oplus$ standing for the Euclidean direct sum. Obviously $\bar{a} = \mathrm{Sc}(a) - \mathrm{Vec}(a)$.

Elements of $\mathbb{R} \subset \mathbb{H}$ and $\mathbb{R}^3 \subset \mathbb{H}$ are termed purely scalar and purely vectorial quaternions respectively.

Let us demonstrate some relations between the decomposition (A1.7) and diverse operations on $\mathbb{H}$.

### A1.6 Proposition

**a)** $a = \bar{a} \iff a \in \mathbb{R}$,

**b)** $\bar{a} = -a \iff a \in \mathbb{R}^3$,

**c)** $a \in \mathbb{R}^3 \iff a^2 \leq 0$.

**A1.7 Proposition** *Addition, multiplication, quaternionic conjugation, and the taking of an inverse are continuous operations with respect to the quaternionic module.*

PROOF. This is a bit of arithmetic in fact. If $a_n \in \mathbf{H} \longrightarrow_{n\to\infty} a \in \mathbb{H}$, $b_n \longrightarrow_{n\to\infty} b$, then

$$|(a+b) - (a_n + b_n)| \leq |a - a_n| + |b - b_n|,$$

$$|ab - a_n b_n| \leq |a| \cdot |b - b_n| + |a - a_n| \cdot b_n,$$

and that is all. Further we have

$$|\bar{a} - \bar{a}_n| = |\overline{a - a_n}| = |a - a_n|.$$

Finally, if $a \neq 0$, then $a_n \neq 0$ for large enough $n$, and for such $n$

$$\begin{aligned} |a^{-1} - a_n^{-1}| &= |a^{-1}(a_n - a) \cdot a_n^{-1}| = \\ &= |a^{-1}| \cdot |a_n - a| \cdot |a_n^{-1}|. \end{aligned}$$

□

**A1.8** Defined as an algebraic object, $\mathbb{H}$ has several nice geometric interpretations and applications. Let $\mathbf{S}^{n-1}$ be the unit sphere in $\mathbb{R}^n$, $n \geq 1$. It is natural to identify $\mathbf{S}^3$ with a set of quaternions of unit modulus:

$$\mathbf{S}^3 = \{x \in \mathbb{H} | |x| = 1\}.$$

$\mathbf{S}^3$, understood in this sense, has the structure of a multiplicative group. In fact,

a) if $\{a, b\} \subset \mathbf{S}^3$ then $|ab| = |a| \cdot |b| = 1 \Rightarrow a \cdot b \in \mathbf{S}^3$;

b) $i_0 \in \mathbf{S}^3$;

c) $a \in \mathbf{S}^3 \Rightarrow \bar{a} \in \mathbf{S}^3$ and $a^{-1} = \bar{a} \in \mathbf{S}^3$.

The group $\mathbf{S}^3$ is non-commutative. It can be considered as an analogue of the group $\mathbf{S}^1$ in $\mathbb{C}$ which is even commutative (with respect to complex number multiplication!). $\mathbf{S}^1$ is also a section of $\mathbf{S}^3$ by a two-dimensional plane $\{x \in \mathbb{R}^4 | x_2 = x_3 = 0\}$. It is curious that a section of $\mathbf{S}^3$ by the hyperplane $\mathbb{R}$; i.e., the unit sphere $\mathbf{S}^2$, is not a group with respect to quaternionic multiplication!

Let us show that for any $a \in \mathbf{S}^3$ there exists $\alpha \in [-\pi, \pi]$ and $b \in \mathbf{S}^2$ such that

$$a = \cos\alpha + b \cdot \sin\alpha. \tag{A1.8}$$

In fact, $a = a_0 + \vec{a} = a_0 + |\vec{a}| \cdot b$ where $b := \frac{1}{|\vec{a}|}\vec{a} \in \mathbb{S}^2$. But $|a_0|^2 + |\vec{a}|^2 = |a|^2 = 1$ which gives the necessary $\alpha$.

For any $a \in \mathbb{H}$ we have obviously:

$$a = |a| \cdot (\cos\alpha + \sin\alpha \cdot b)$$

with $\alpha$ and $b$ as above.

**A1.9 Proposition** *For any $a \in \mathbb{S}^3$ define a mapping $f_a$ by the rule*

$$f_a : x \in \mathbb{H} \longmapsto axa^{-1} \in \mathbb{H}.$$

*Then:*

**a)** *$f_a$ is multiplicative and additive, i.e.,*

$$f_a(xy) = f_a(x)f_a(y);$$

$$f_a(x+y) = f_a(x) + f_a(y);$$

**b)** *$f_a$ is $\mathbb{R}$-homogeneous, i.e., $\forall \lambda \in \mathbb{R}$*

$$f_a(\lambda x) = \lambda f_a(x);$$

**c)** *$f_a$ is an isometric automorphism of a skew field $\mathbb{H}$;*

**d)** *the scalar product in $\mathbb{R}^4$ is invariant under $f_a : \forall x, y \in \mathbb{H}$*

$$< f_a(x), f_a(y) >_{\mathbb{R}^4} = < x, y >_{\mathbb{R}^4}.$$

**e)** *$f_a \cdot f_b = f_{ab}$.*

PROOF.

**a)** $f_a(xy) = axya^{-1} = axa^{-1} \cdot aya^{-1} = f_a(x) \cdot f_a(y)$.

**b)** Obvious.

**c)** Defining

$$f_a^{-1} : y \in \mathbb{H} \longmapsto a^{-1}ya$$

we see that it is an inverse to $f_a$ and it has the same properties. Further,

$$|f_a(x)| = |axa^{-1}| = |a| \cdot |x| \cdot |a|^{-1} = |a|.$$

**d)** We have

$$\begin{aligned} < x, y >_{\mathbb{R}^4} &:= \textstyle\sum_{k=0}^{3} x_k \cdot y_k \\ &= \mathrm{Sc}(\bar{x}y) = \mathrm{Sc}(\bar{y}x) = \tfrac{1}{2}(\bar{x}y + \bar{y}x). \end{aligned}$$

Hence

$$\begin{aligned} < f_a(x), f_a(y) >_{\mathbb{R}^4} &= \tfrac{1}{2}(\overline{f_a(x)} \cdot f_a(y) + \overline{f_a(y)} \cdot f_a(x)) = \\ &= \tfrac{1}{2}(\overline{axa^{-1}} \cdot aya^{-1} + \overline{aya^{-1}} \cdot axa^{-1}) = \\ &= \tfrac{1}{2}(\bar{a}^{-1}\bar{x}\bar{a}aya^{-1} + \bar{a}^{-1}\bar{y}\bar{a} \cdot axa^{-1}) = \\ &= \tfrac{1}{2}(\bar{x}y + \bar{y}x)\bar{a}^{-1} \cdot a^{-1} = \tfrac{1}{2}(\bar{x}y + \bar{y}x) =< x, y >_{\mathbb{R}^4} . \end{aligned}$$

**e)** $\forall x \in \mathbb{H} \quad f_a f_b(x) = a(bxb^{-1})a^{-1} = ab \cdot x \cdot (ab)^{-1} = f_{ab}(x)$.

□

**A1.10 Corollary** *The mapping $f_a$ has the following properties:*

**a)** *It preserves the orthogonality in the sense of $\mathbb{R}^4$; i.e., if $< x, y >_{\mathbb{R}^4} = 0$, then*

$$< f_a(x), f_a(y) >_{\mathbb{R}^4} = 0.$$

b) *The restriction* $f_a|\mathbb{R}$ *of* $f_a$ *onto* $\mathbb{R}$ *is the identity mapping; the restriction* $f_a|\mathbb{R}^3$ *is an automorphism.*

c) $f_a(x) = x_0 + f_a(\vec{x})$.

### A1.11 Corollary

a) $f_a$ *determines a rotation of the Euclidean space* $\mathbb{R}^4$.

b) $f_a|\mathbb{R}^3$ *determines a rotation of the Euclidean space* $\mathbb{R}^3$.

PROOF.

a) By A1.8 $a = \cos\alpha + \sin\alpha \cdot c$. Let $t$ run over the whole segment $[0;1]$, and define a mapping

$$a_t : t \in [0;1] \longmapsto \cos\alpha t + \sin\alpha t \cdot c \in \mathbb{S}^3.$$

In particular, $a_0 = i_0$, $a_1 = a$, and $a_t$ is continuous. Hence $f_{a_t}$ connects continuously the identity mapping (which is a rotation) and $f_a$.

b) Follows from what we have just proved and Corollary A1.10 b).

□

**A1.12** The above corollary says that quaternionic multiplication gives a compact and elegant description of (at least some) rotations in $\mathbb{R}^3$ and $\mathbb{R}^4$. We continue the examination of the relations between rotations in $\mathbb{R}^3$ and $\mathbb{R}^4$, and quaternions. Let, as always, $O(\mathbb{R}^n) = O_n(\mathbb{R})$ denote the group of all orthogonal transformations (= motions) in $\mathbb{R}^n$, and let $O^+(\mathbb{R}^n) = O_n^+(\mathbb{R})$ be its subgroup of rotations.

It is quite remarkable that while mappings of the form $f_a$ describe only part of $O_4^+(\mathbb{R})$, they exhaust the whole of $O_3^+(\mathbb{R})$ (it is curious to compare this with $\mathbb{R}^2 = \mathbb{C}$ where all rotations are multiplications by elements of $\mathbb{S}^1$). To prove this is our nearest aim. Firstly let us state both theorems.

**A1.13 Theorem** (Rotations in $\mathbb{R}^3$ and quaternions) *The sets $O_3^+(\mathbb{R})$ and $\{f_a|\mathbb{R}^3|a \in \mathbb{S}^3\}$ coincide, i.e., any rotation of $\mathbb{R}^3$ is of the form $f_a$. Also, if $\{a,b\} \subset \mathbb{S}^3$ then*

$$f_a = f_b \iff a = \pm b.$$

For this reason quaternions from $\mathbb{S}^3$ are sometimes called "versors" or "rotators".

**A1.14 Theorem** (Rotations in $\mathbb{R}^4$ and quaternions) *Let $f_{a,b} : x \in \mathbb{H} \longmapsto axb^{-1}$ for $\{a,b\} \in \mathbb{S}^3$. The sets $O_4^+(\mathbb{R})$ and $\{f_{a,b}|a \in \mathbb{S}^3, b \in \mathbb{S}^3\}$ coincide, i.e., any $f_{a,b}$ is a rotation and there are no others. Also, if $f_{a,b} = f_{A,B}$ then $a = \pm A$, $b = \pm B$ with both plus or both minus.*

Both proofs are anticipated by a series of preliminary results interesting in themselves.

**A1.15 Observations**

1) Let $\psi_1, \psi_2$ be two purely vectorial quaternions, i.e., two three-dimensional vectors. Assume also that they are unit and orthogonal in the sense of $\mathbb{R}^3$.

   Let $\psi_3$ be their quaternionic product: $\psi_3 := \psi_1 \cdot \psi_2$. Then it is easily seen that $\vec{\psi}_3$ is a vector product of $\psi_1$ and $\psi_2$, that is $\vec{\psi}_3 = \vec{\psi}_1 \times \vec{\psi}_2$, and thus a triplet $\psi = \{\psi_1, \psi_2, \psi_3\}$ possesses exactly the same properties as $\psi_{st} = \{i_1, i_2, i_3\}$.

2) The transformation $f_a|\mathbb{R}^3$ depends on three arbitrary parameters. The transformation $f_{a,b}$ depends on six arbitrary parameters. Theorems A1.13 and A1.14 explain the geometrical meaning of these parameters.

**A1.16 Proposition** *Let $\psi_1, \psi_2, \psi_3$ be as in Observation A1.15 and let $P$ denote a plane generated by $\vec{\psi}_2$, $\vec{\psi}_3$. Let $a := \cos\alpha + \sin\alpha \cdot \psi_1$. Then the rotation $f_a$ defined in A1.9 is a rotation of $P$ around the axis $\psi_1$ by the angle $2|\alpha|$, from $\psi_2$ to $\psi_3$ for $\alpha > 0$, and from $\psi_3$ to $\psi_2$ for $\alpha < 0$.*

PROOF. Let us verify that $f_a(\psi_1) = \psi_1$. In fact,

$$\begin{aligned} f_a(\psi_1) = a\psi_1 a^{-1} &= (\cos\alpha + \sin\alpha \cdot \psi_1)\psi_1(\cos\alpha - \sin\alpha \cdot \psi_1) = \\ &= \psi_1(\cos\alpha + \sin\alpha \cdot \psi_1)(\cos\alpha - \sin\alpha \cdot \psi_1) = \psi_1. \end{aligned}$$

Analogously

$$f_a(\psi_2) = \cos 2\alpha \cdot \psi_2 + \sin 2\alpha \cdot \psi_3,$$

$$f_a(\psi_3) = \cos 2\alpha \cdot \psi_2 - \sin 2\alpha \cdot \psi_3,$$

which is just what is required. □

**A1.17 Proposition** *Let $\varphi, \psi$ be two purely vectorial quaternions of unit modulus. Then there exists a rotation $f_a$ which maps $\varphi$ to $\psi$, i.e., $\exists a \in \mathbb{S}^3$ such that $f_a(\varphi) = \psi$. If additionally $\varphi$ tends to $\psi$ then $\alpha \to 0$ and $a \to 1$.*

PROOF. One can consider $\varphi, \psi$ as two points of the unit sphere $\mathbb{S}^2$ in $\mathbb{R}^3$, and thus geometrically it is obvious that $\mathbb{S}^2$ can be rotated such that $\varphi$ goes to $\psi$. The fine point is whether this rotation can be represented as $f_a$.

Construct such an $f_a$. Let $P$ be the plane pulled on $\varphi, \psi$. We take $\psi_2, \psi_3$ to be purely vectorial quaternions lying on $P$ and of unit modulus, and let $\psi_1 := \psi_2 \cdot \psi_3$.

Let $2\alpha$ be the angle between $\varphi$ and $\psi$; then $f_a$ with $a := \cos\alpha + \sin\alpha \cdot \varphi$ gives all we need. □

**A1.18 Lemma**

1) *Let $\psi_1, \psi_2, \psi_3 := \psi_1 \cdot \psi_2$ be as above. Then there exists a rotation $f_c$ such that*

$$f_c(i_s) = \psi_s, s \in \mathbb{N}_3,$$

*and if $\psi_s \to \varphi_s$ then $c \to 1$ and $f_c \to I$, the identity transformation of $\mathbb{R}^3$.*

2) *Let $\varphi_1, \varphi_2, \varphi_3$ and $\psi_1, \psi_2, \psi_3$ be two orthonormal bases in $\mathbb{R}^3$. Then there exists a rotation $f_c$ such that $f_c(\psi_s) = \varphi_s$, $s \in \mathbb{N}_3$. If besides $\psi_s \to \varphi_s$ $\forall s$, then $c \to 1$, $f_c \to I$.*

PROOF.

1) By the last proposition $\exists f_a : f_a(i_1) = \psi_1$. Further, if $\psi_s \to i_s$ $\forall s$ then $a \to 1$. Let $P$ be the plane generated by $i_2, i_3$, and let $P' := f_a(P)$. Then $P'$ contains $f_a(i_2), f_a(i_3)$, $\psi_2, \psi_3$. Now there exists $f_b$, $b := \sin\beta + \cos\beta \cdot \psi_1$, such that $f_a(i_2)$ goes to $\psi_2$. Besides if $\psi_s \to i_s$ $\forall s$ then $b \to 1$. Thus we have $f_c(f_a(i_2)) = \psi_2$. Let $c := b \cdot a$; then $f_c(i_1) = \psi_1$, $f_c(i_2) = \psi_2$. But also $f_c(i_3) = f_c(i_1 \cdot i_2) = f_c(i_1) \cdot f_c(i_2) = \psi_1 \cdot \psi_2 = \psi_3$.

2) Denote $\psi_3' = \psi_1 \cdot \psi_2$. Then $\psi_3' = \varepsilon_0 \psi_3$ with $\varepsilon_0 = \pm 1$. The same holds for $\varphi_3' := \varphi_1 \cdot \varphi_2$, that is $\varphi_3' = \varepsilon_1 \cdot \varphi_3$ with $\varepsilon_1 = \pm 1$. When $\psi_s \to \varphi_s$ then $\psi_3' \to \varphi_3'$, thus $\varepsilon_0 = \varepsilon_1 =: \varepsilon = \pm 1$. Consider now two orthonormalized bases $\psi_1, \psi_2, \varepsilon\psi_3$ and $\varphi_1, \varphi_2, \varepsilon\varphi_3$. Applying the first part of the lemma we arrive at the necessary result.

□

**A1.19** Proof of Theorem A1.13. We lack proving that for any $g \in O_3^+(\mathbb{R})$ there exists $a \in \mathbb{S}^3$ such that $g(x) = f_a(x)$, $\forall x \in \mathbb{R}^3$.

It is known from linear algebra that any $g \in O_3^+(\mathbb{R})$ can be connected with the identity rotation $I$ with a continuous curve lying strictly inside of $O_3^+(\mathbb{R})$; i.e., $\exists \varphi_t : t \in [0,1] \longmapsto O_3^+(\mathbb{R})$ such that $\varphi \in C([0;1], O_3^+(\mathbb{R}))$, $\varphi_0 = I$, $\varphi_1 = g$. Let $t_0 = 0 < t_1 < \ldots < t_n = 1$ be any partition chosen such that $\varphi_{t_1}$ is close to $\varphi_0$ (in an obvious sense), $\varphi_{t_2}$ is close to $\varphi_{t_1}$, etc. Take now $\varphi_{t_1}(i_1)$, $\varphi_{t_1}(i_2)$, $\varphi_{t_1}(i_3)$. By Lemma A1.18 there exists $c_1 \in \mathbb{S}^3$ such that $f_{c_1}(i_s) = \varphi_{t_1}(i_s)$ $\forall s \in \mathbb{N}_3$. Repeat this procedure for $\varphi_{t_2}(\varphi_{t_1}))$, $\varphi_{t_2}(\varphi_{t_1}(i_2))$, $\varphi_{t_2}(\varphi_{t_1}(i_3))$ and get $f_{c_2}$. Finally we get $f_{c_n}$ such that

$$g = f_{c_n} \circ f_{c_{n-1}} \circ \ldots \circ f_{c_1} = f_{c_n \cdot c_{n-1} \cdots c_1}.$$

Thus $a := c_n c_{n-1} \ldots c_1$ is what we are looking for.

Let now $a, b$ be of $\mathbb{S}^3$. If $a = \pm b$ then it is obvious that $f_a = f_b$. To prove the opposite assertion consider the rotation $f_c := f_{a^{-1}} \cdot f_b$. Then $c = a^{-1} b$. Since $f_a = f_b$

we get $f_c = I$. Let $c = \cos\alpha + \sin\alpha \cdot u$, then $f_c$ is a rotation around the axis $u$ by the angle $2\alpha$. But such a rotation is identical if and only if $2\alpha = 2q\pi \iff \alpha = q\pi$ which implies that $c = \pm 1$ and $a = \pm b$.

**A1.20** Proof of Theorem A1.14.

1) First let us show that $f_{a,b}$ is a rotation. Obviously $|f_{a,b}(x)| = |x|$, and it is not difficult to obtain that $f_{a,b}$ preserves the scalar product in $\mathbb{R}^4$ also. Thus $f_{a,b}$ is an automorphism of the Euclidean space $\mathbb{R}^4$. Now we demonstrate that there exists a continuous path in $O_4^+(\mathbb{R})$ connecting $f_{a,b}$ with $I$. For this, write:

$$a = \cos\alpha + \sin\alpha \cdot u,$$

$$b = \cos\beta + \sin\beta \cdot v,$$

with $u, v, \alpha, \beta$ as above. Also as above, one can verify that, for $t \in [0, 1]$ and

$$a_t = \cos\alpha t + \sin\alpha t \cdot u,$$

$$b_t = \cos\beta t + \sin\beta t \cdot v,$$

the path we are looking for is given by $f_{a_t,b_t}$.

2) Now let $g \in O_4^+(\mathbb{R})$. Denote $\varepsilon := g(i_0) \in \mathbb{S}^3$ and $g' := \varepsilon^{-1}g$. $g'$ is a composition of two rotations and thus also a rotation.

But $g'|\mathbb{R} = I_{\mathbb{R}}$, hence $g'|\mathbb{R}^3$ is an automorphism of $\mathbb{R}^3$ which implies that $g'|\mathbb{R}^3$ is a rotation in $\mathbb{R}^3$. By Theorem A1.13, $\exists c \in \mathbb{S}^3$ such that

$$g'(\vec{x}) = c\vec{x}c^{-1} \quad \forall \vec{x} \in \mathbb{R}^3.$$

Again taking into account that $g'|\mathbb{R} = I_{\mathbb{R}}$ we get: $\forall x \in \mathbb{R}^4$

$$\begin{aligned} g'(x) &= g'(x_0 + \vec{x}) = g'(x_0) + g'(\vec{x}) = \\ &= x_0 + c\vec{x}c^{-1} = c(x_0 + \vec{x})c^{-1}, \end{aligned}$$

and thus

$$g(x) = \varepsilon \cdot c \cdot x \cdot c^{-1} =: axb^{-1}.$$

If now $g_1 = g_2 \in O_4^+(\mathbb{R})$ then $g_1(i_0) = \varepsilon_1$ and $g_2(i_0) = \varepsilon_1$, thus $g_1' = g_2'$. By Theorem A1.13, $c_1$ and $c_2$ generated by them can differ in their signs only which gives the last statement.

**A1.21 Corollary** (Non-rotational automorphism and quaternions) *There exist automorphisms of $\mathbb{R}^3$ which are not rotations.*

In fact, let $g \in O_3(\mathbb{R})$ be defined on the canonical basis as follows:

$$g(i_1) = i_1, \qquad g(i_2) = i_2, \qquad g(i_3) = -i_3.$$

Suppose that $g = f_a$ for some $a \in \mathbb{S}^3$. Then

$$-i_3 = f_a(i_3) = f_a(i_1 i_2) = f_a(i_1) f_a(i_2) = i_1 i_2 = i_3$$

which is a contradiction.

Note that $g = Z_3|\mathbb{R}^3$, the restriction of the quaternionic conjugation with respect to $i_3$ onto $\mathbb{R}$. Thus we have shown that the automorphism $Z_3$ has quite another nature, not that of multiplication by quaternions.

The facts proved admit some other useful reformulations and interpretations. A part of them is given below.

**A1.22 Theorem** (Skew-field automorphisms of $\mathbb{H}$) *All automorphisms of a skew-field $\mathbb{H}$ are of the form*

$$\mathcal{F} : x \in \mathbb{H} \longmapsto x_0 + g(\vec{x})$$

*where $g \in O_3^+(\mathbb{R})$; i.e., $\exists a \in \mathbb{S}^3$ such that*

$$\mathcal{F} = I_{\mathbb{R}} \oplus f_a.$$

*In particular, this means that all automorphisms of* $\mathbb{H}$ *are internal* (recall that in each associative algebra with identity an invertible element generates an algebra automorphism of the form $x \longmapsto axa^{-1}$ which is just called internal).

The proof is an easy consequence of what was said above. But the fact itself has a quite profound nature: it reflects an essential difference between $\mathbb{H}$ and its "generator", the space $\mathbb{R}^4$: in the latter all the four coordinate axes are completely equivalent, while in $\mathbb{H}$ the difference between $i_0$ and each imaginary unit $i_1, i_2, i_3$ implies that the coordinate axes $x_1, x_2, x_3$ become abruptly different (geometrically!) from $x_0$.

In other words, having introduced one more (multiplicative) structure in $\mathbb{R}^4$ one is led to the asymmetry of some properties.

Note also that each automorphism $\mathcal{F}$ of $\mathbb{H}$ is a proper orthogonal transformation (= rotation) of $\mathbb{R}^4$, that is $\mathcal{F} \in O_4^+(\mathbb{R})$.

**A1.23 Theorem** (Quaternionic multiplication and group homomorphisms)

1) *Let* $a \in \mathbf{S}^3$, $f_a(x) := axa^{-1}$; *then the mapping*

$$f : a \longmapsto f_a \in O_3^+(\mathbb{R})$$

*is a continuous and surjective group homomorphism with* $\ker f = \{\pm i_0\}$.

2) *Let* $\{a, b\} \subset \mathbf{S}^3$, $f_{a,b}(x) := axb^{-1}$; *then the mapping*

$$f : (a, b) \in \mathbf{S}^3 \times \mathbf{S}^3 \longmapsto f_{a,b} \in O_4^+(\mathbb{R})$$

*is a continuous and surjective group homomorphism with*

$$\ker f = \{(i_0, i_0), (-i_0, -i_0)\}.$$

The proof follows directly from Theorem A1.13 and Theorem A1.14.

**A1.24 Corollary** *Let* $\Lambda := \{I; -I\} \subset O_4^+(\mathbb{R})$ *be a cyclic group of order 2,* $PO_4^+(\mathbb{R}) := O_4^+/\Lambda$, *a quotient group. Then*

$$PO_4^+(\mathbb{R}) = O_3^+(\mathbb{R}) \times O_3^+(\mathbb{R}).$$

It appeared that just this property distinguishes the group $O_4^+(\mathbb{R})$ in the family $O_n^+(\mathbb{R})$ for $n \geq 3$.

**A1.25** Here we demonstrate one more relation between rotations in $\mathbb{R}^3$ and quaternions. Let $\vec{x} \in \mathbb{R}^3$. We want to rotate it by the angle $\varphi$ around the axis orthogonal to $\vec{x}$ and determined by the unit vector $\vec{p}$.

Any $a \in \mathbf{S}^3$, by (A1.8), is of the form $a = \cos\alpha + b\sin\alpha$ with $b \in \mathbf{S}^2$. Take $\alpha = \varphi$ and $b = \vec{p}$. Then:

$$\begin{aligned} \mathrm{Sc}(a \cdot x) &= a_0 \cdot x_0 - <\vec{a}, \vec{x}>_{\mathbb{R}^3} = \\ &= \cos\varphi \cdot 0 - \sin\varphi \cdot <\vec{x}, \vec{p}> = 0, \end{aligned}$$

$$\begin{aligned} \mathrm{Vec}(a \cdot x) &= \cos\varphi \cdot x + 0 \cdot p \cdot \sin\varphi + \sin\varphi \cdot [\vec{p} \times \vec{x}] =: \\ &=: x \cdot \varphi + y \cdot \sin\varphi \end{aligned}$$

where $y \in \mathbb{R}^3$, $|y| = |p| \cdot |x| \cdot \sin(\widehat{p, x}) = |x|$, $\vec{y} \perp \vec{x}$. Hence

$$a \cdot x = x \cdot \cos\varphi + y \cdot \sin\varphi,$$

that is, to make the rotation we are looking for, it is enough to multiply the vector $\vec{x}$ by quaternionic rules by the quaternion $a := \cos\varphi + p \cdot \sin\varphi$, which gives a quite simple and elegant way to obtain arithmetically the resulting vector.

**A1.26** The same idea of the usefulness of quaternionic arithmetic for describing rotations can be illustrated by solving the problem of the composition of two rotations. Assume that we want to rotate the space $\mathbb{R}^3$ first by an angle $2\varphi_1$ around the axis determined by the unit vector $\vec{P}_1$, then to rotate it again with parameters $2\varphi_2$ and

$\vec{P}_2$. By Theorem A1.13, $\varphi_1, \vec{P}_1$ and $\varphi_2, \vec{P}_2$ determine unit quaternions $a_1$ and $a_2$ for which there exists a unit vector $p$ and an angle $\varphi$ such that

$$a_1 \cdot a_2 = \cos\varphi + p\sin\varphi,$$

i.e., the resulting rotation is obtained. It is quite difficult to solve this problem without quaternions.

Of course this means that a composition of rotations is a simple geometric equivalent of the product of two quaternions.

By the way, the non-commutativity of quaternions corresponds to the well-known fact that changing the order of two rotations around the same point we arrive at, generally speaking, different results.

**A1.27** Let $g$ be any rotation in $\mathbb{R}^3$ and $\lambda$ be any real constant. Then $\lambda \cdot g : x \longmapsto \lambda g(x)$ can be naturally called "a rotational dilatation" of $\mathbb{R}^3$. It is quite clear how to describe all of them in quaternionic terms. The same holds in $\mathbb{R}^4$.

**A1.28** We now give several more remarks and observations. Let $q$ be a quaternion. Then it satisfies the equality

$$q^2 - (q + \bar{q}) \cdot q + q \cdot \bar{q} = 0,$$

or equivalently

$$q^2 - 2q_0 \cdot q + |q|^2 = 0.$$

This means that the equation

$$x^2 - 2a \cdot x + b^2 = 0,$$

with $a \in \mathbb{R}$, $b \in \mathbb{R}$, has, in general, infinitely many quaternionic solutions: any $q \in \mathbb{H}$ with $q_0 = a, |q| = |b|$ solves it. In particular, as we know already the set of solutions to the equation $x^2 + 1 = 0$ coincides with the unit sphere in $\mathbb{R}^3$.

**A1.29** Here let $\{a_{11}, a_{12}, a_{21}, a_{22}\} \subset \mathbb{H}$. What can be said about the invertibility of the matrix

$$A = \begin{pmatrix} a_{11} & a_{12} \\ a_{21} & a_{22} \end{pmatrix},$$

i.e., whether there exists

$$X = \begin{pmatrix} x_{11} & x_{12} \\ x_{21} & x_{22} \end{pmatrix}$$

with entries from $\mathbb{H}$ such that

$$AX = XA = E_2 := \begin{pmatrix} 1 & 0 \\ 0 & 1 \end{pmatrix}.$$

The problem is equivalent to solving the system

$$\begin{cases} a_{11}x_{11} + a_{12}x_{21} = 1, \\ a_{11}x_{12} + a_{12}x_{22} = 0, \\ a_{21}x_{11} + a_{22}x_{21} = 0, \\ a_{21}x_{12} + a_{22}x_{22} = 1. \end{cases}$$

If $a_{11} = a_{12} = a_{21} = a_{22} = 0$ then $A$ is obviously non-invertible. Hence $A$ can be invertible only if at least one of the entries differs from zero.

More exactly, $A$ is invertible if and only if one of the following four symmetric conditions is satisfied:

1) $a_{11} \neq 0$ and $\Delta_{11} := a_{21} \cdot a_{11}^{-1}a_{12} - a_{22} \neq 0$. Then

$$A^{-1} = \begin{pmatrix} a_{11}^{-1}(1 + a_{12}\Delta_{11}^{-1} \cdot a_{21}a_{11}^{-1}); & -a_{11}^{-1}a_{12}\Delta_{11}^{-1} \\ -\Delta_{11}^{-1}a_{21}a_{11}^{-1}; & \Delta_{11}^{-1} \end{pmatrix}.$$

2) $a_{12} \neq 0$ and $\Delta_{12} := a_{22}a_{12}^{-1}a_{11} - a_{21} \neq 0$. Then

$$A^{-1} = \begin{pmatrix} -\Delta_{12}^{-1} a_{22} a_{12}^{-1}; & -\Delta_{12}^{-1} \\ a_{12}^{-1}(1 - a_{11}\Delta_{12}^{-1} a_{22} a_{12}^{-1}); & a_{12}^{-1} a_{11} \Delta_{12}^{-1} \end{pmatrix}.$$

3) $a_{21} \neq 0$ and $\Delta_{21} := a_{11} a_{21}^{-1} a_{22} - a_{12} \neq 0$. Then

$$A^{-1} = \begin{pmatrix} a_{21}^{-1} a_{22} \Delta_{21}^{-1}; & a_{21}^{-1}(1 - a_{22}\Delta_{21}^{-1} a_{11} a_{21}^{-1}) \\ -\Delta_{21}^{-1}; & \Delta_{21}^{-1} a_{11} a_{21}^{-1} \end{pmatrix}.$$

4) $a_{22} \neq 0$ and $\Delta_{22} := -a_{11} + a_{12} a_{22}^{-1} a_{21} \neq 0$. Then

$$A^{-1} = \begin{pmatrix} -\Delta_{22}^{-1}; & \Delta_{22}^{-1} a_{12} a_{22}^{-1} \\ a_{22}^{-1} a_{21} \Delta_{22}^{-1}; & a_{22}^{-1}(1 - a_{21}\Delta_{22}^{-1} a_{12} a_{22}^{-1}) \end{pmatrix}.$$

These formulas can be simplified a little. Denote $a_{12}^{-1} - a_{22}^{-1} a_{21} =: \sigma_1$, $a_{21}^{-1} a_{22} - a_{11}^{-1} a_{12} =: \sigma_2$ when they exist. Then

$$\Delta_{11} = a_{21}\sigma_2,$$

$$\Delta_{22} = a_{12}\sigma_1,$$

$$\Delta_{12} = a_{22}\sigma_1,$$

$$\Delta_{21} = a_{11}\sigma_2.$$

One of these exists and is invertible iff $\sigma_1 \neq 0$ or $\sigma_2 \neq 0$, i.e., $\Delta_{11}$ and $\Delta_{21}$ are simultaneously invertible, $\Delta_{22}$ and $\Delta_{12}$ are simultaneously invertible. The inverses then are:

I) $a_{11} \neq 0$, $\sigma_2 \neq 0$,

$$A^{-1} = \begin{pmatrix} a_{11}^{-1}(1 + a_{12}\sigma_2^{-1} a_{11}^{-1}); & -a_{11}^{-1} a_{12} \sigma_2^{-1} a_{12}^{-1} \\ -\sigma_2^{-1} a_{11}^{-1}; & \sigma_2^{-1} a_{21}^{-1} \end{pmatrix},$$

**II)** $a_{12} \neq 0$, $\sigma_1 \neq 0$,

$$A^{-1} = \begin{pmatrix} \sigma_1^{-1} a_{12}^{-1}; & -\sigma_1^{-1} a_{22}^{-1} \\ a_{12}^{-1}(1 - a_{11}\sigma_1^{-1} a_{12}^{-1}); & a_{12}^{-1} a_{11} \sigma_1^{-1} a_{22}^{-1} \end{pmatrix},$$

**III)** $a_{21} \neq 0$, $\sigma_2 \neq 0$,

$$A^{-1} = \begin{pmatrix} a_{21}^{-1} a_{22} \sigma_2^{-1} a_{11}^{-1}; & a_{21}^{-1}(1 - a_{22}\sigma_2^{-1} a_{21}^{-1}) \\ -\sigma_2^{-1} a_{11}^{-1}; & -\sigma_2^{-1} a_{21}^{-1} \end{pmatrix},$$

**IV)** $a_{22} \neq 0$, $\sigma_1 \neq 0$,

$$A^{-1} = \begin{pmatrix} -\sigma_1^{-1} a_{12}^{-1}; & \sigma_1^{-1} a_{22}^{-1} \\ a_{22}^{-1} a_{21} \sigma_1^{-1} a_{12}^{-1}; & a_{22}^{-1}(1 - a_{21}\sigma_1^{-1} a_{22}^{-1}) \end{pmatrix}.$$

# 2 Representations of real quaternions

**A2.1** In Section A1, above, we were using, almost always, a canonical way of introducing the real quaternions as elements of $\mathbb{R}^4$. Now we describe several other representations of $\mathbb{H}$. By "representation" we mean here not only representation as a homomorphism into some algebra but another form of writing elements from $\mathbb{H}$ which shows how they are related to other important mathematical objects.

First of all, recall that in Subsection A1.5 we obtained in fact that each quaternion $a$ can be written as a sum (in the sense of addition in $\mathbb{H}$) of scalar and vectorial quaternions. Initially scalars and vectors are of a different nature and cannot be added. Of course scalars can be identified with one-dimensional vectors which, in turn, are particular cases of three-dimensional ones, but this isn't what we want to say: quaternionic addition just permits us to add scalar and vectors without losing their proper natures. This would not be of great interest if $\mathbb{H}$ did not have a good multiplication also.

Let $a, b \in \mathbb{H}$, $a = a_0 + \vec{a}$, $b = b_0 + \vec{b}$. Then the quaternionic product $a \cdot b$ is given by (A1.3). Let us present it as follows:

$$\begin{aligned} a \cdot b &= (a_0 b_0 - \textstyle\sum_{k=1}^3 a_k b_k) i_0 + \\ &+ a_0 \textstyle\sum_{k=1}^3 b_k i_k + b_0 \sum_{k=1}^3 a_k i_k + \\ &+ \left( \begin{vmatrix} a_2 & a_3 \\ b_2 & b_3 \end{vmatrix} i_1 + \begin{vmatrix} a_3 & a_1 \\ b_3 & b_1 \end{vmatrix} i_2 + \begin{vmatrix} a_1 & a_2 \\ b_1 & b_2 \end{vmatrix} i_3 \right). \end{aligned}$$

Recalling the scalar and vector products in $\mathbb{R}^3$:

$$< \vec{a}, \vec{b} > := \sum_{k=1}^{3} a_k b_k;$$

$$[\vec{a}, \vec{b}] := \begin{vmatrix} i_1 & i_2 & i_3 \\ a_1 & a_2 & a_3 \\ b_1 & b_2 & b_3 \end{vmatrix},$$

we arrive at the formula

$$a \cdot b = a_0 b_0 - < \vec{a}, \vec{b} > + a_0 \vec{b} + b_0 \vec{a} + [\vec{a}, \vec{b}]. \qquad \text{(A2.1)}$$

On the left we have the quaternionic multiplication of two arbitrary quaternions while on the right we see a series of multiplications: that of two scalars; multiplications of vectors by scalars, and two types of product of vectors. In particular (A2.1) says that

$$\mathrm{Sc}(a \cdot b) = \mathrm{Sc}(a) \cdot \mathrm{Sc}(b) - < \mathrm{Vec}(a); \mathrm{Vec}(b) >;$$

$$\mathrm{Vec}(a \cdot b) = \mathrm{Sc}(a) \cdot \mathrm{Vec}(b) + \mathrm{Sc}(b) \cdot \mathrm{Vec}(a) + [\mathrm{Vec}(a); \mathrm{Vec}(b)].$$

The most impressive form (A2.1) takes if $a_0 = b_0 = 0$:

$$\vec{a}\cdot\vec{b} = -<\vec{a},\vec{b}> + [\vec{a},\vec{b}]. \tag{A2.2}$$

In some sense (A2.2) explains why the scalar product and the vector (= cross) product of three-dimensional vectors lack almost all good arithmetical properties of numbers, real and complex: both of them are parts, fragments of the "numerical" product, that of quaternions which does have good arithmetical properties.

(A2.1) together with (A2.2) allow us to rewrite all quaternionic facts in vectorial form thus obtaining pithy assertions. Just a few examples:

**A2.2 Proposition** (Vector and scalar products of vectors, and quaternionic multiplication) *Let* $a, b \in \mathbb{R}^3$. *Then*

$$\begin{aligned} <\vec{a},\vec{b}> &= -\tfrac{1}{2}(\vec{a}\cdot\vec{b} + \vec{b}\cdot\vec{a}), \\ [\vec{a}\times\vec{b}] &= \tfrac{1}{2}(\vec{a}\cdot\vec{b} - \vec{b}\cdot\vec{a}). \end{aligned} \tag{A2.3}$$

PROOF. From (A2.2)

$$\begin{aligned} \vec{b}\cdot\vec{a} &= -<\vec{a},\vec{b}> + [\vec{b},\vec{a}] = \\ &= -<\vec{a},\vec{b}> - [\vec{a},\vec{b}]. \end{aligned} \tag{A2.4}$$

Adding and subtracting (A2.2) and (A2.4) we get (A2.3). □

**A2.3 Proposition** (Orthogonality and collineation of vectors in quaternionic terms) *Let* $\vec{a},\vec{b} \in \mathbb{R}^3$. *Then*

$$\begin{aligned} \vec{a}\perp\vec{b} &\iff \vec{a}\cdot\vec{b} + \vec{b}\cdot\vec{a} = 0; \\ \vec{a}||\vec{b} &\iff \vec{a}\cdot\vec{b} - \vec{b}\cdot\vec{a} = 0. \end{aligned} \tag{A2.5}$$

PROOF. This follows directly from (A2.3). □

**A2.4 Proposition** (Double-vector and mixed products of vectors, and quaternionic multiplication) *Let* $\{\vec{a},\vec{b},\vec{c}\} \subset \mathbb{R}^3$. *Then*

$$[\vec{a}\times\vec{b}]\times\vec{c}=\frac{1}{4}(\vec{a}\cdot\vec{b}\cdot\vec{c}+\vec{c}\cdot\vec{b}\cdot\vec{a}-\vec{b}\cdot\vec{a}\cdot\vec{c}-\vec{c}\cdot\vec{a}\cdot\vec{b}) \tag{A2.6}$$

$$\begin{aligned}(\vec{a};\vec{b};\vec{c}) &:= <[\vec{a}\times\vec{b}],\vec{c}> \\ &= \tfrac{1}{4}(\vec{b}\cdot\vec{a}\cdot\vec{c}+\vec{c}\cdot\vec{b}\cdot\vec{a}-\vec{a}\cdot\vec{b}\cdot\vec{c}-\vec{c}\cdot\vec{a}\cdot\vec{b}).\end{aligned} \tag{A2.7}$$

**Proof.** By (A2.3) applied twice we have

$$\begin{aligned}[\vec{a}\times\vec{b}]\times\vec{c} &= \tfrac{1}{2}([\vec{a}\times\vec{b}]\cdot\vec{c}-\vec{c}\cdot[\vec{a}\times\vec{b}]) = \\ &= \tfrac{1}{4}((\vec{a}\cdot\vec{b}-\vec{b}\cdot\vec{a})\cdot\vec{c}-\vec{c}\cdot(\vec{a}\cdot\vec{b}-\vec{b}\cdot\vec{a})) \\ &= \tfrac{1}{4}(\vec{a}\cdot\vec{b}\cdot\vec{c}-\vec{b}\cdot\vec{a}\cdot\vec{c}-\vec{c}\cdot\vec{a}\cdot\vec{b}+\vec{c}\cdot\vec{b}\cdot\vec{a}).\end{aligned}$$

Analogously for (A2.7). □

**A2.5 Proposition** (Properties of the vector product via quaternionic operations) *Let* $\{\vec{a},\vec{b},\vec{c}\}\subset\mathbb{R}^3$. *Then*

1) $[\vec{a}\times\vec{a}]=0$,

2) $[(\vec{a}+\vec{b})\times\vec{c}]=[\vec{a}\times\vec{c}]+[\vec{b}\times\vec{c}]$,

3) $\forall\lambda\in\mathbb{R}\ \ [\lambda\vec{a}\times\vec{b}]=[\vec{a}\times\lambda\vec{b}]=\lambda[\vec{a}\times\vec{b}]$.

**Proof.** This follows directly from (A2.3). □

**A2.6 Proposition** (Vector operations and associativity of quaternions) *Let* $\{\vec{a},\vec{b},\vec{c}\}\subset\mathbb{R}^3$. *Then*

1)

$$(\vec{a};\vec{b};\vec{c})=(\vec{b};\vec{c};\vec{a})=(\vec{c};\vec{a};\vec{b}); \tag{A2.8}$$

2) *if two vectors are parallel (in particular, equal) then their mixed product is zero;*

3)

$$\vec{a}\times[\vec{b}\times\vec{c}]-\vec{a}\cdot<\vec{b},\vec{c}>=[\vec{a}\times\vec{b}]\times\vec{c}-<\vec{a},\vec{b}>\vec{c}. \tag{A2.9}$$

PROOF. Considering $\vec{a}, \vec{b}, \vec{c}$ as quaternions we have

$$\vec{a} \cdot (\vec{b} \cdot \vec{c}) = (\vec{a} \cdot \vec{b}) \cdot \vec{c}, \tag{A2.10}$$

or

$$\vec{a} \cdot (- < \vec{b}, \vec{c} > + [\vec{b}, \vec{c}]) = (- < \vec{a}, \vec{b} > + [\vec{a} \times \vec{b}]) \cdot \vec{c},$$

$$-\vec{a} \cdot < \vec{b}, \vec{c} > - < \vec{a}, [\vec{b} \times \vec{c}] > + [\vec{a} \times [\vec{b} \times \vec{c}]] = < \vec{a}, \vec{b} > \cdot \vec{c} - < [\vec{a} \times \vec{b}], \vec{c} > + [[\vec{a} \times \vec{b}] \times \vec{c}]. \tag{A2.11}$$

This means that (A2.8) and (A2.9) are nothing more than the scalar and vectorial parts of the quaternionic equality (A2.11). We should mention here that, for instance, (A2.9) is not too easy to prove if one wants to use vector tools only.

Let now $\vec{b} = \lambda \vec{a}$, $\lambda \in \mathbb{R}$. Then by (A2.7)

$$(\vec{a}; \vec{b}; \vec{c}) = \lambda(\vec{a}^2 \cdot \vec{c} + \vec{c} \cdot \vec{a}^2 - \vec{a}^2 \cdot \vec{c} - \vec{c} \cdot \vec{a}^2) = 0.$$

□

**A2.7** Continuing in this way we can obtain many other formulas of vectorial arithmetic. But this is not our goal here, and we proceed to the next representation of real quaternions.

Let $\{a, b\} \subset \mathbb{H}$. Then their product is given by (A1.3). The right-hand side of (A1.3) can be written as follows:

$$\begin{pmatrix} a_0 & -a_1 & -a_2 & -a_3 \\ a_1 & a_0 & -a_3 & a_2 \\ a_2 & a_3 & a_0 & -a_1 \\ a_3 & -a_2 & a_1 & a_0 \end{pmatrix} \cdot \begin{pmatrix} b_0 \\ b_1 \\ b_2 \\ b_3 \end{pmatrix},$$

which means that the element of $\mathbb{R}^4$, obtained from $b \in \mathbb{R}^4$ by multiplying "quaternionically" by $a \in \mathbb{R}^4$, coincides with the result of multiplying $b$ by the matrix

$$B_l(a) := \begin{pmatrix} a_0 & -a_1 & -a_2 & -a_3 \\ a_1 & a_0 & -a_3 & a_2 \\ a_2 & a_3 & a_0 & -a_1 \\ a_3 & -a_2 & a_1 & a_0 \end{pmatrix}. \tag{A2.12}$$

Quite analogously, considering $b \cdot a$ we get

$$B_r(a) := \begin{pmatrix} a_0 & -a_1 & -a_2 & -a_3 \\ a_1 & a_0 & a_3 & -a_2 \\ a_2 & -a_3 & a_0 & a_1 \\ a_3 & a_2 & -a_1 & a_0 \end{pmatrix}. \tag{A2.13}$$

For an arbitrary $a \in \mathbb{H}$, $B_l(a)$ and $B_r(a)$ are called its left and right regular representations respectively. These terms are taken from the algebra where the mode of their introduction described above has a deep sense and motivation. However we are going to use this mode as a leading reason to get the sets

$$\mathcal{B}_l := \{B_l(a) | \forall a \in \mathbb{H}\}, \quad \mathcal{B}_r := \{B_r(a) | \forall a \in \mathbb{H}\}. \tag{A2.14}$$

**A2.8** What is the exact relation between $\mathcal{B}_l$ and $\mathbb{H}$? One can compute directly that:

1) $B_l(a+b) = B_l(a) + B_l(b)$;

2) $B_l(a \cdot b) = B_l(a) \circ B_l(b)$, where "$\circ$" denotes the usual matricial multiplication;

3) in particular, $\forall \lambda \in \mathbb{R}$, $B_l(\lambda a) = \lambda B_l(a)$.

The second property means that the product of any two matrices from $\mathcal{B}_l$ again belongs to $\mathcal{B}_l$, whence all three say that a mapping

$$a \in \mathbb{H} \longmapsto B_l(a) \in \mathbb{R}^{4\times 4}$$

($\mathbb{R}^{4\times 4}$ being the set of real $4 \times 4$ matrices) is a homomorphism of (real) algebras. Moreover,

4) $B_l(i_0) = E_4$, the unit $4 \times 4$ matrix;

5) $B_l(\bar{a}) = B_l^T(a)$, the transposed matrix;

6) $\det B_l(a) = (\sum_{k=0}^{3} a_k^2)^2 = |a|^4$.

Thus, $B_l$ is an algebra with a unit and is invariant with respect to the operation of transposition. The sixth property means that any non-zero matrix from $B_l$ is invertible. Hence $B_l$ is a (skew) field. Combining all six properties we see that $\mathbb{H}$, as a real skew field, is isomorphic to $B_l$.

The same remains true for $B_r$, the only difference being that instead of 2) one can prove 2′) $B_r(a \cdot b) = B_r(b) \circ B_r(a)$.

**A2.9** Both the definition of quaternions and the representations obtained (vectorial and matricial) demonstrate a "real" aspect of their nature. It appears that the set $\mathbb{H}$ can be modelled quite successfully and fruitfully using the concept of a complex number.

Take $a \in \mathbb{H}$; i.e., $a = a_0 + a_1 i_1 + a_2 i_2 + a_3 i_3$, then

$$\begin{aligned} a &= (a_0 + a_1 i_1) + a_2 i_2 + a_3 i_1 i_2 = \\ &= (a_0 + a_1 i_1) + (a_2 + a_3 i_1) \cdot i_2 =: z_1 + z_2 i_2 \end{aligned}$$

with $z_1 := a_0 + a_1 i_1$; $z_2 := a_2 + a_3 i_1$. It is obvious that the set $\mathbb{C}(i_1) := \{a + b i_1 | a, b \in \mathbb{R}\}$ is a "copy" of the set $\mathbb{C}$ of complex numbers; i.e., $\mathbb{C}(i_1)$ and $\mathbb{C}$ are isometrically isomorphic. Thus we have

$$\mathbb{H} = \mathbb{C}(i_1) \pm \mathbb{C}(i_1) \cdot i_2 \cong \mathbb{C}^2. \tag{A2.15}$$

In other words quaternionic multiplication allows us to define a "good" multiplicative structure in the (left) complex linear space $\mathbb{C}^2$, transforming it to a complex (over $\mathbb{C}(i_1)$) algebra.

Let $\tilde{\mathbb{C}}^2$ denote such an algebra; we will describe what is induced in it from $\mathbb{H}$. We denote in this section $i_1 =: i$, $i_2 =: j$ thus writing any element $z \in \tilde{\mathbb{C}}^2$ as a "complex number" $z = z_1 + z_2 j$ whose components $z_1, z_2$ are also complex numbers but with respect to another imaginary unit: $z_k := x_k + iy_k, \{x_k, y_k\} \subset \mathbb{R}$, $i^2 = -1$, $j^2 = -1$, and $i \cdot j = -j \cdot i$. First of all, the interrelation between complex numbers and $j$ should be defined as follows:

$$z_k \cdot j =: j z_k^*, \tag{A2.16}$$

$z_k^* = x_k - iy_k$; i.e., having defined initially the product of $z_k$ by $j$ on the left, we define now multiplication on the left also. (A2.16) has two immediate consequences. One of them requires us to define conjugation on $\tilde{\mathbb{C}}^2$ by the formula

$$\bar{z} = \overline{z_1 + z_2 j} := z_1^* - z_2 j. \tag{A2.17}$$

The second one requires us to define a multiplication in $\tilde{\mathbb{C}}^2$ by the formula

$$z \cdot w := (z_1 w_1 - z_2 w_2^*) + (z_1 w_2 + z_2 w_1^*) j. \tag{A2.18}$$

One illustration of (A2.15) and (A2.16):

$$\begin{aligned} \overline{z \cdot w} &= (z_1 w_1 - z_2 w_2^*)^* - (z_1 w_2 + z_2 w_1^*) j = \\ &= (z_1^* \cdot w_1^* - z_2^* \cdot w_2) - (z_1 w_2 + z_2 w_1^*) j = \\ &= (w_1^* - w_2 j)(z_1^* - z_2 j) = \bar{w} \cdot \bar{z}. \end{aligned}$$

We have also

$$\begin{aligned} |z|^2 &:= z \cdot \bar{z} = (z_1 + z_2 j)(z_1^* - z_2 j) = \\ &= z_1 \cdot z_1^* + z_2 z_2^* + (-z_1 z_2 + z_2 z_1) j = |z_1|^2 + |z_2|^2 \end{aligned}$$

with $|z_k|^2 := x_k^2 + y_k^2$, the usual complex number modulus, which means, of course, that any non-zero element $z$ of $\tilde{\mathbb{C}}^2$ is invertible, and its inverse

$$z^{-1} = \frac{z_1^* - z_2 j}{|z|^2}.$$

This idea can be clearly developed quite substantially.

**A2.10** More often one can meet a "complex number model" of real quaternions in matrix form. Moreover, sometimes (real) quaternions are introduced just as $2 \times 2$ matrices with complex number entries.

Consider the set $\tilde{\mathbb{H}}$ of all matrices of the form

$$q = \begin{pmatrix} z & w \\ -w^* & z^* \end{pmatrix} \tag{A2.19}$$

where $z, w$ are "usual" complex numbers and $z^*$ denotes now the usual complex conjugation. (To understand better how the matrix (A2.19) arises one can repeat the trick of Subsection A2.7 changing a quaternionic product of $c \in \mathbb{H}$ by $a = z + wj$ on the left to a product of a matrix and a complex two-dimensional vector.) The following properties can be obtained by a straightforward calculation:

1) $\tilde{\mathbb{H}}$ is an additive abelian group;

2) $\tilde{\mathbb{H}}\backslash\{0\}$ is a non-abelian group with respect to matrix multiplication;

3) $\tilde{\mathbb{H}}$ is a $\mathbb{R}$-four-dimensional vector space with a basis

$$\mathbf{I} := \begin{pmatrix} 1 & 0 \\ 0 & 1 \end{pmatrix}, \mathbf{I}_1 := \begin{pmatrix} i & 0 \\ 0 & -i \end{pmatrix}, \mathbf{I}_2 := \begin{pmatrix} 0 & 1 \\ -1 & 0 \end{pmatrix}, \mathbf{I}_3 := \begin{pmatrix} 0 & i \\ i & 0 \end{pmatrix}.$$

In particular, $\mathbf{I}_1 \cdot \mathbf{I}_2 = \mathbf{I}_3$, etc. This 4-tuple generates a multiplicative group of order 8.

4) The centre of the group $\tilde{\mathbb{H}}\backslash\mathbb{O}$ is $\mathbb{R} \cdot \mathbf{I}$. Hence $\tilde{\mathbb{H}}$ and $\mathbb{H}$ are isomorphic as skew fields. In other words every real quaternion can be identified with a matrix (A2.19), and one can work with quaternions as with complex matrices of special form but using usual arithmetic operations over matrices.

   Note also that

$$\det q = z \cdot z^* + ww^* = |z|^2 + |w|^2 = |z + wj|^2. \tag{A2.20}$$

**A2.11** There exists an intimate relation between quaternions, as matrices (A2.19), and unitary matrices. Recall that a $2 \times 2$ complex matrix $A$ is called unitary if $A \circ A^\dagger = A^\dagger \circ A = E_2$, $E_2$ being the identity matrix, $A^\dagger := (A^*)^T = (A^T)^*$ the adjoint matrix. Obviously, $\det A^\dagger \circ A = 1 \Rightarrow |\det A|^2 = 1$.

The set of such matrices is denoted by $\mathcal{U}(2)$, and its subgroup of matrices with $\det A = 1$ is denoted by $\mathcal{SU}(2)$. Also $SL(2;\mathbb{C})$ denotes all invertible matrices with determinant one, and for any $A, B \in \mathbb{C}^{2\times 2}$ a complex-valued scalar product is defined by

$$[A; B] := tr A \circ B^\dagger = a_{11}b_{11}^* + a_{12}b_{12}^* + a_{21}b_{21}^* + a_{22}b_{22}^*,$$

and

$$||A|| := \sqrt{[A; A]}.$$

**A2.12 Proposition** (Unitary matrices and quaternions)

1) $\mathcal{SU}(2) = \tilde{\mathbb{H}} \cap \mathbb{S}^3$.

2) $\forall A \in \mathcal{SU}(2)$, $||A||^2 = 2$.

PROOF. If $A \in \mathcal{U}(2)$ then $|a_{11}|^2+|a_{12}|^2 = 1$, $|a_{21}|^2+|a_{22}|^2 = 1$, $a_{11}\cdot a_{21}^*+a_{12}\cdot a_{22}^* = 0$. Its subgroup $\mathcal{SU}(2)$ is singled out by the additional condition $a_{11}a_{22} - a_{12}a_{21} = 1$. Altogether, this gives exactly $\tilde{\mathbb{H}} \cap \mathbb{S}^3$. Also $||A||^2 = (|z|^2 + |w|^2) \cdot 2 = 2$. □

**A2.13** Let us summarize here what a real quaternion is.

1) It is a four-component vector: $a = \sum_{k=0}^{3} a_k i_k$.

2) It is a formal sum of a scalar and a three-dimensional vector $a = a_0 + \vec{a}$.

3) It is a real $4 \times 4$ matrix:

$$a = \begin{pmatrix} a_0 & -a_1 & -a_2 & -a_3 \\ a_1 & a_0 & -a_3 & a_2 \\ a_2 & a_3 & a_0 & -a_1 \\ a_3 & -a_2 & a_1 & a_0 \end{pmatrix} \text{ or } a = \begin{pmatrix} a_0 & -a_1 & -a_2 & -a_3 \\ a_1 & a_0 & a_3 & -a_2 \\ a_2 & -a_3 & a_0 & a_1 \\ a_3 & a_2 & -a_1 & a_0 \end{pmatrix}.$$

**4)** It is an ordered pair of two complex numbers:

$$a = z_1 + z_2 j.$$

**5)** It is a complex $2 \times 2$ matrix:

$$q = \begin{pmatrix} z & w \\ -w^* & z^* \end{pmatrix}.$$

# 3 Complex quaternions

**A3.1** In introducing real quaternions; i.e., in fact, a multiplicative structure in $\mathbb{R}^4$, the main point is the multiplicative rules for the canonical basis given by (A1.1). But a collection of 4-vectors $i_0 = (1,0,0,0)$, $i_1 = (0,1,0,0)$, $i_2 = (0,0,1,0)$, $i_3 = (0,0,0,1)$ may be considered as a basis in the complex vector space $\mathbb{C}^4$, $\mathbb{C} = \mathbb{C}(i)$. After having assumed the latter one can begin to follow the line described in Appendix 1 with minor changes. First of all, how should we multiply $i_k$ and $i$, the imaginary unit of $\mathbb{C}$? Of course, we should have $i_0 \cdot i = i \cdot i_0$. Furthermore, $\forall k \in \mathbb{N}_3$ by definition

$$i_k \cdot i = i \cdot i_k. \tag{A3.1}$$

Thus we now have $a = \sum_{k=0}^{3} a_k i_k$ with $\{a_k\} \subset \mathbb{C}$. The set $\{a = \sum_{k=0}^{3} a_k i_k\} =: \mathbb{H}(\mathbb{C})$ is a set of complex quaternions, or biquaternions. Operations in $\mathbb{H}(\mathbb{C})$ are determined by the formulas from Subsection A1.2, in particular two complex quaternions $a$ and $b$ are multiplied by (A1.3) where of course all expressions in parentheses are complex numbers. These operations allow us to represent each complex quaternion in the form

$$a = a^{(1)} + a^{(2)} \cdot i = a^{(1)} + ia^{(2)} \tag{A3.2}$$

where $a^{(1)}, a^{(2)} \in \mathbb{H}$. (A3.2) says that a complex quaternion is a kind of "complex number" but with real-quaternionic components. One can say that complex quaternions combine properties of both real quaternions and usual complex numbers.

**A3.2** In particular $\mathbb{H}(\mathbb{C})$ inherits conjugations both from $\mathbb{H}$ and from $\mathbb{C}$:

$$\begin{aligned}
Z_1 &: a \in \mathbb{H}(\mathbb{C}) \longmapsto a_0 - a_1 i_1 + a_2 i_2 + a_3 i_3, \\
Z_2 &: a \in \mathbb{H}(\mathbb{C}) \longmapsto a_0 + a_1 i_1 - a_2 i_2 + a_3 i_3, \\
Z_3 &: a \in \mathbb{H}(\mathbb{C}) \longmapsto a_0 + a_1 i_1 + a_2 i_2 - a_3 i_3, \\
Z &: a \longmapsto \bar{a} := a_0 - a_1 i_1 - a_2 i_2 - a_3 i_3 = \overline{a^{(1)}} + i \cdot \overline{a^{(2)}}, \\
Z_{\mathbb{C}} &: a \in \mathbb{H}(\mathbb{C}) \longmapsto a^* := \textstyle\sum_{k=0}^{3} a_k^* \cdot i_k = a^{(1)} - ia^{(2)},
\end{aligned}$$

where $a_k^* := \mathrm{Re} a_k - i \mathrm{Im} a_k$ is complex conjugation. In other words $Z_k$ affects the imaginary unit $i_k$ only, $Z$ affects all three quaternionic imaginary units, and $Z_{\mathbb{C}}$ affects the complex-number imaginary unit. $Z_k$ and $Z$ clearly enjoy the properties:

1) $Z_k(\lambda a + b) = \lambda Z_k(a) + Z_k(b)$, $\forall \lambda \in \mathbb{C}$, $\forall a, b \in \mathbb{H}(\mathbb{C})$,

2) $Z_k(a \cdot b) = Z_k(a) \cdot Z_k(b)$,

3) $Z(a \cdot b) = \overline{a \cdot b} = \bar{b} \cdot \bar{a}$,

4) $\bar{\bar{a}} = a$,

5) $\overline{a + b} = \bar{a} + \bar{b}$.

As for the complex conjugation $Z_{\mathbb{C}}$ in $\mathbb{H}(\mathbb{C})$ one can easily calculate:

6) $(\alpha(a + b)\beta)^* = \alpha(a^* + b^*)\beta$, for $\alpha, \beta \in \mathbb{H}$ and $a, b \in \mathbb{H}(\mathbb{C})$,

7) $(a^*)^* = a$,

8) $(a \cdot b)^* = a^* \cdot b^*$.

**A3.3** What are the analogues of (A1.5) now; i.e., what is obtained by multiplying two conjugate (in different meanings) complex quaternions? We have

$$\begin{aligned} a \cdot a^* &= ((a^{(1)} + ia^{(2)}) \cdot (a^{(1)} - ia^{(2)}) = \\ &= (a^{(1)})^2 + (a^{(2)})^2 + i(a^{(2)}a^{(1)} - a^{(1)}a^{(2)}) \neq \\ &\neq a^* \cdot a = a^*(a^*)^* = (a \cdot a^*)^*, \end{aligned} \tag{A3.3}$$

$$\bar{a} \cdot a = a \cdot \bar{a} = \sum_{k=0}^{3} a_k^2 = |a^{(1)}|^2 - |a^{(2)}|^2 + 2i < a^{(1)}, a^{(2)} >_{\mathbb{R}^4} . \tag{A3.4}$$

(A3.3) says that, generally speaking, $a \cdot a^*$ is a complex quaternion, whilst (A3.4) determines a product of two quaternionically conjugate complex quaternions as a complex number. In general,

$$a \cdot \bar{a} \neq |a|^2_{\mathbb{R}^8} = |a^{(1)}|^2 + |a^{(2)}|^2. \tag{A3.5}$$

Equality (A3.4) is related to Frobenius' theorem mentioned in Subsection A1.2: the algebra $\mathbb{H}(\mathbb{C})$ does have zero-divisors, and (A3.4) explains why. In fact, if $a \cdot \bar{a} \neq 0$, then

$$a^{-1} := \frac{1}{a \cdot \bar{a}} \cdot \bar{a} \tag{A3.6}$$

determines an inverse to $a \in \mathbb{H}(\mathbb{C})$. If $a \cdot \bar{a} = 0$ for $a \neq 0$ then $a$ is obviously a zero-divisor in $\mathbb{H}(\mathbb{C})$. The set $\mathfrak{S}$ of all zero-divisors is described in Lemma 1.7 which can be complemented with a certain geometric interpretation:

$$a = a^{(1)} + ia^{(2)} \in \mathfrak{S} \iff |a^{(1)}| = |a^{(2)}| \text{ and } a^{(1)} \perp_{\mathbb{R}^4} a^{(2)}.$$

**A3.4** We will not go into depth in the theory of complex quaternions as we have for real ones. It is clear that some of these properties can be trivially generalized, for example, the relations between the multiplication of complex quaternions and the scalar and vector products of complex three-dimensional vectors: in fact, all formulas remain true. It is clear also that there are properties which do have their analogues in $\mathbb{H}(\mathbb{C})$ but not too trivially. So we limit ourselves to a brief description of representations of complex quaternions (= different models of $\mathbb{H}(\mathbb{C})$). What we already have is:

1) A complex quaternion is a complex four-component vector $a = \sum_{k=0}^{3} a_k i_k$.

2) A complex quaternion is a pair of real quaternions:

$$a = a^{(1)} + ia^{(2)} = \text{Re}a + i\text{Im}a.$$

Of course we obviously get that

3) A complex quaternion is a formal sum of a complex scalar and a complex three-dimensional vector:

$$a = a_0 + \vec{a}.$$

Reasoning as in Subsection A2.7 we find that

4) A complex quaternion is a $4 \times 4$ matrix with complex entries:

$$a = \begin{pmatrix} a_0 & -a_1 & -a_2 & -a_3 \\ a_1 & a_0 & -a_3 & a_2 \\ a_2 & a_3 & a_0 & -a_1 \\ a_3 & -a_2 & a_1 & a_0 \end{pmatrix} \text{ or } a = \begin{pmatrix} a_0 & -a_1 & -a_2 & -a_3 \\ a_1 & a_0 & a_3 & -a_2 \\ a_2 & -a_3 & a_0 & a_1 \\ a_3 & a_2 & -a_1 & a_0 \end{pmatrix}.$$

Having transformed $a = \sum_{k=0}^{3} a_k i_k$ to the expression $a = (a_0 + a_1 i_1) + (a_2 + a_3 i_1) i_2$ we concluded that real quaternions turn $\mathbb{C}^2$ into an algebra. If $a \in \mathbb{H}(\mathbb{C})$ then a new structure arises, that of bicomplex numbers: $a_0 + a_1 i_1$ may be considered as a "complex" number whose "real" and "imaginary" parts are some other "complex numbers" but in such a way that both imaginary units commute. The latter shows the great difference with real quaternions where both imaginary units anticommute. Both bicomplex numbers and real quaternions are algebras generated by $\mathbb{R}^4$ but the second one is a skew field while the first has zero divisors. Hence:

5) A complex quaternion is an ordered pair of bicomplex numbers $z_1, z_2$ :

$$a = z_1 + z_2 \cdot i_2.$$

This implies immediately:

6) A complex quaternion is a $2 \times 2$ matrix with bicomplex entries:

$$q = \begin{pmatrix} z_1 & z_2 \\ -z_2^* & z_1^* \end{pmatrix}.$$

7) The last representation of $\mathbb{H}(\mathbb{C})$ which has been actively used is that of the Dirac algebra. Let $\gamma_0, \gamma_1, \gamma_2, \gamma_3$ be the standard Dirac matrices, see Subsection 1.4. Then each complex quaternion $b = \sum_{k=0}^{3} b_k i_k$ is a linear combination of products of the Dirac matrices:

$$\begin{aligned} b &= \text{Re}b_0 \cdot E_4 + \text{Re}b_1 \cdot \gamma_3\gamma_2 + \text{Re}b_2 \cdot \gamma_1\gamma_3 + \text{Re}b_3 \cdot \gamma_1\gamma_2 + \\ &+ \text{Im}b_0 \cdot \gamma_0\gamma_1\gamma_2\gamma_3 + \text{Im}b_1 \cdot \gamma_0\gamma_1 + \text{Im}b_2 \cdot \gamma_0\gamma_2 + \text{Im}b_3 \cdot \gamma_0\gamma_3. \end{aligned}$$

**A3.5** Just to give an illustration of an analogue of the relation between real quaternions and rotations of the Euclidean space $\mathbb{R}^4$, let us consider a relation between the Lorentz transformation and complex quaternions. This transformation can be treated as a rotation of a non-Euclidean four-dimensional space and as such can be conveniently represented with the aid of complex quaternions.

Let now $\mathbb{R}^4$ denote the arithmetic four-dimensional real space. A matrix $A \in \mathbb{R}^{4\times4}$ is said to be a Lorentz transformation if for $y, x \in \mathbb{R}^4$, $y = Ax$ there holds:

$$\sum_{k=1}^{3} x_k^2 - c^2 x_4^2 = \sum_{k=1}^{3} y_k^2 - c^2 y_4^2, \tag{A3.7}$$

i.e., $A$ preserves pseudo-metric $\sum_{k=1}^{3} x_k^2 - c^2 x_4^2$ in $\mathbb{R}^4$, hence $A$ is an analogue of the motion. Introduce the following complex quaternions:

$$\begin{aligned} q &= \textstyle\sum_{k=1}^{3} x_k i_k + icx_4, \\ p &= \textstyle\sum_{k=1}^{3} y_k i_k + icy_4. \end{aligned} \tag{A3.8}$$

For them $qq^* = -(\sum_{k=1}^{3} x_k^2 - c^2 x_4^2)$, $pp^* = -(\sum_{k=1}^{3} y_k^2 - c^2 y_4^2)$, and hence one can prove, in exact analogy to the real-quaternionic case, that the formula

$$p = \frac{a}{(a \cdot a^*)} \cdot q \cdot \frac{b}{(b \cdot b^*)} \tag{A3.9}$$

describes a linear transformation with the property (A3.7) if $a$ and $b$ are complex quaternions with $aa^* \neq 0$, $bb^* \neq 0$.

It is necessary sometimes to get real coefficients with $a_{44} > 0$. These can be obtained as follows. Let $\alpha, \beta \in \mathbb{R}^4$ with $\alpha \perp_{\mathbb{R}^4} \beta, |\alpha| > |\beta|$. Define

$$\begin{aligned} a &:= i_1(\alpha_1 + i\beta_1) + i_2(\alpha_2 + i\beta_2) + i_3(\alpha_3 + i\beta_3) + (\alpha_4 + i\beta_4), \\ b &:= (\alpha_4 - i\beta_4) - i_1(\alpha_1 - i\beta_1) - i_2(\alpha_2 - i\beta_2) - i_3(\alpha_3 - i\beta_3) = \bar{a}^*, \\ M &:= |\alpha|^2 - |\beta|^2. \end{aligned} \tag{A3.10}$$

The formula

$$p = \frac{a \cdot q \cdot b}{M} \tag{A3.11}$$

together with (A3.10) gives all Lorentz' transformations.

# 4 $\alpha$-holomorphic functions of two real variables

**A4.1** In Chapter 1 we constructed a theory of $\alpha$-holomorphic functions of three real variables. The present appendix is aimed at the construction of an analogous theory for functions of two real variables. There are at least two reasons to do this. First of all, some phenomena described by hyperholomorphic functions are of sufficient interest for the plane situation which requires exact formulas, theorems, and definitions. Secondly, although the general approach to the study of the two-dimensional situation is quite clear, nevertheless there exist enough obstacles to the rigorous mathematical treatment of the problem to justify such this work. By the way, only the existence of a good multiplicative structure (that of complex numbers) in the plane separates the three-dimensional case and the two-dimensional one, making the latter interesting in itself. The reader can find some speculations on this point below. It is worth mentioning that till now quite a lot of articles have been published devoted to the two-dimensional Helmholtz operator, hence it can be expected that the two-dimensional version of our theory will find its own applications and usage (some recent works on the two-dimensional Helmholtz equation are [20, 25, 41, 99]). One more reason is that one can consider the two-dimensional case as a good model for constructing the Clifford analysis related to the Helmholtz equation with a complex or Clifford wave number. For the pioneering works in that direction we refer the reader to [131, 59, 84, 19, 12]. Below we give a very brief sketch of some definitions and some peculiarities of the case under consideration, based on a detailed description given in [109, 110]. These articles, following the general line described in Chapter 1 of this book, give all proofs directly, without referring to the three-dimensional case. It is possible but not convenient and not usually easier to extract from the three-dimensional theory what we need for the two-dimensional situation.

**A4.2** We consider $\mathbb{H}(\mathbb{C})$-valued functions defined in a domain $\Omega \subset \mathbb{R}^2$. On the left $\mathbb{H}(\mathbb{C})$-module $C^2(\Omega; \mathbb{H}(\mathbb{C}))$ we introduce the two-dimensional Helmholtz operator with a quaternionic wave number $\lambda$: $\Delta_\lambda := \Delta_{\mathbb{R}^2} + M^\lambda$, where $\Delta_{\mathbb{R}^2} := \partial_1^2 + \partial_2^2$,

$\partial_k := \frac{\partial}{\partial x_k}$. Let $\psi := \{\psi^1, \psi^2\} \subset \mathbb{H}(\mathbb{R}) \times \mathbb{H}(\mathbb{R})$, and denote $\bar{\psi} := \{\overline{\psi^1}, \overline{\psi^2}\}$. Let

$$\partial_\psi := \psi^1 \partial_1 + \psi^2 \partial_2. \tag{A4.1}$$

As above, the equalities

$$\partial_\psi \cdot \partial_{\bar{\psi}} = \partial_{\bar{\psi}} \cdot \partial_\psi = \Delta_{\mathbb{R}^2} \tag{A4.2}$$

hold if, and only if,

$$\psi^j \cdot \overline{\psi^k} + \psi^k \cdot \overline{\psi^j} = 2\delta_{jk},$$

for any $j, k$ from $\mathbb{N}_2$, $\delta_{jk}$ being the Kronecker delta. For purely vectorial $\psi$ the factorization (A4.2) becomes

$$\partial_\psi^2 = -\Delta_{\mathbb{R}^2}, \tag{A4.3}$$

which is paradoxically different, for many purposes, from the factorization

$$(\partial_1 + i\partial_2)(\partial_1 - i\partial_2) = \Delta_{\mathbb{R}^2} \tag{A4.4}$$

by the usual complex Cauchy–Riemann operators

$$\overline{\partial} := \partial_1 + i\partial_2, \qquad \partial := \partial_1 - i\partial_2. \tag{A4.5}$$

Let $\alpha^2 = \lambda$, $\alpha \in \mathbb{H}(\mathbb{C})$, and

$$\partial_{\psi,\alpha} := \partial_\psi + M^\alpha.$$

Then (A4.2) implies the following factorizations of the two-dimensional Helmholtz operator:

$$\begin{aligned} \Delta_\lambda &= \partial_{\bar{\psi},\alpha} \cdot \partial_{\psi,\alpha} = \partial_{\psi,\alpha} \cdot \partial_{\bar{\psi},\alpha} = \\ &= -\partial_{\psi,-\alpha} \cdot \partial_{\psi,\alpha} = -\partial_{\psi,\alpha} \cdot \partial_{\psi,-\alpha}. \end{aligned}$$

Thus an $\alpha$-holomorphic function is any solution of the equation

$$\partial_{\psi,\alpha} f := \partial_\psi f + f\alpha = 0. \tag{A4.6}$$

Consider again the operator (A4.1). Observe that we can write $\partial_\psi = \psi^1(\partial_1 + \psi^3\partial_2)$, with $\psi^3 := \overline{\psi^1} \cdot \psi^2$. In general, $\psi^3 \cdot \psi^3 \neq -1$, and thus the operator $\partial_1 + \psi^3\partial_2$ only formally looks like (A4.5). But if $\psi$ is a structural set with $\psi^1$, $\psi^2$ purely vectorial then $\psi^3$ does have the property $\psi^3 \cdot \psi^3 = -1$ and thus can play the role of an imaginary unit. Hence $\ker \partial_\psi = \ker(\partial_1 + \psi^3\partial_2)$, and we get some kind of holomorphy with respect to $\partial_1 + \psi^3\partial_2$. A fine point arises here. The functions under consideration are complex-valued but a complex unit $i$ does not coincide with the "complex unit" $\psi^3$ in $\partial_1 + \psi^3\partial_2$! Thus we have, generally speaking, something different from the "usual" holomorphy. But for an $\mathbb{H}(\mathbb{R})$- valued $f$ we are, in fact, in the "normal" holomorphic situation for $\mathbb{C}^2$-valued functions of one complex variable $x + \psi^3 y$.

Furthermore this opens the way to constructing a theory of $\alpha$-holomorphic complex functions (i.e., those depending on a complex variable and with complex values).

**A4.3** It is well known that if $\lambda = \alpha^2 \in \mathbb{C}$, a fundamental solution $\theta_\alpha$ of $\Delta_\lambda$ is given by

$$\theta_\alpha[u] := \frac{-i}{4} H_0^1(\alpha|u|),$$

where $u = (x, y)$ and $H_0^1$ is one of the Hankel functions, namely, the Hankel function of the first kind of order zero. Their properties can be found elsewhere. For our purposes we need immediately that $H_0^1$ satisfies the relation

$$\frac{\partial}{\partial u} H_0^1(u) = -H_1^1(u),$$

with $H_1^1$ the Hankel function of the first kind of order one.

Like the three-dimensional case this generates fundamental solutions both for the two-dimensional Helmholtz operator and for the Cauchy–Riemann operator.

**A4.4 Theorem** *Let* $\lambda \in \mathbb{H}(\mathbb{C})$. *Then a fundamental solution* $\theta_\alpha$ *for the operator* $\Delta_{\alpha^2} = \Delta + M^{\alpha^2}$ *when* $u \in \mathbb{R}^2 \setminus \{0\}$ *is given by:*

1) *If* $\alpha^2 \notin \mathfrak{S}$, $\vec{\alpha}^2 \neq 0$, *let* $\gamma := \sqrt{\vec{\alpha}^2} \in \mathbb{C}$ *and* $\xi_\pm := \alpha_0 \pm \gamma$, *then*

$$\begin{aligned}\theta_\alpha(u) &= \tfrac{1}{2\gamma}(\theta_{\xi_+}(u)(\gamma+\vec{\alpha})+\theta_{\xi_-}(u)(\gamma-\vec{\alpha})) = \\ &= \tfrac{-i}{8\gamma}[H_0^1((\alpha_0+\gamma)|u|)(\gamma+\vec{\alpha})+H_0^1((\alpha_0-\gamma)|u|)(\gamma-\vec{\alpha})].\end{aligned}$$

2) *If* $\alpha^2 \notin \mathfrak{G}$, $\vec{\alpha}^2 = 0$ *then*

$$\begin{aligned}\theta_\alpha(u) &= \theta_{\alpha_0}(u)+\tfrac{\partial}{\partial\alpha_0}[\theta_{\alpha_0}(u)]\vec{\alpha} = \\ &= \tfrac{-i}{4}[H_0^1(\alpha_0|u|)-H_1^1(\alpha_0|(u)|)|u|\vec{\alpha}].\end{aligned}$$

3) *If* $\alpha^2 \in \mathfrak{G}$ *then*

$$\theta_\alpha(u)=\frac{1}{2\alpha_0}(\theta_{2\alpha_0}(u)\cdot\alpha+\theta(u)\cdot\vec{\alpha}) = -\frac{i}{8\alpha_0}[H_0^1(2\alpha_0|u|)\cdot\alpha+\frac{1}{2\alpha_0}\theta(u)\cdot\vec{\alpha}].$$

**A4.5 Theorem** (**Fundamental solution for the operator** $\partial_{\psi,\alpha}$) *Let* $\alpha \in \mathbb{H}(\mathbb{C})$, $u = (x,y) \in \mathbb{R}^2\setminus\{0\}$. *Then a fundamental solution* $\mathcal{K}_{\psi,\alpha}$ *for the operator* $\partial_{\psi,\alpha}$ *is given by the formulas*

1) *If* $\alpha = \alpha_0 \in \mathbb{C}$, *then*

$$\mathcal{K}_{\psi,\alpha}(u) = -\frac{i}{4}\alpha[H_1^1(\alpha|u|)\frac{u_\psi}{|u|}-H_0^1(\alpha|u|)]$$

*where* $u_\psi = \psi^1 x + \psi^2 y$.

2) *If* $\alpha \notin \mathfrak{G}$, $\vec{\alpha}^2 \neq 0$, *then*

$$\mathcal{K}_{\psi,\alpha}(u) = P^+[\mathcal{K}_{\psi,\xi_+}](u)+P^-[\mathcal{K}_{\psi,\xi_-}](u),$$

*where* $\mathcal{K}_{\psi,\xi_\pm}(u)$ *are the above-defined fundamental solutions to the operators* $\partial_{\psi,\xi_\pm}$ *with the complex parameters* $\xi_\pm$.

3) *If* $\alpha \notin \mathfrak{G}$, $\vec{\alpha}^2 = 0$, *then*

$$\mathcal{K}_{\psi,\alpha} = \mathcal{K}_{\psi,\alpha_0} + \frac{\partial}{\partial \alpha_0}[\mathcal{K}_{\psi,\alpha_0}]\vec{\alpha}.$$

4) *If* $\alpha \in \mathfrak{G}$, $\alpha_0 \neq 0$, *then*

$$\mathcal{K}_{\psi,\alpha} = P^+[\mathcal{K}_{\psi,2\alpha_0}] + P^-[\mathcal{K}_{\psi,0}].$$

5) *If* $\alpha \in \mathfrak{G}$, $\alpha_0 = 0$, *then*

$$\mathcal{K}_{\psi,\alpha} = \mathcal{K}_{\psi,0} + \theta \cdot \alpha = \mathcal{K}_{\psi,0} + \theta \cdot \alpha.$$

**A4.6** Here we describe the precise relations between metaharmonic functions of two real variables and the respective $\alpha$-holomorphic functions (compare with Theorem 3.3).

Let $\lambda \notin \mathfrak{G} \cup \{0\}$ and let $f$ be $\lambda$-metaharmonic. Then for any structural set $\psi$ with purely vectorial $\psi^1$, $\psi^2$ and for any complex-quaternionic square root $\alpha$ of $\lambda$, there exist (uniquely) two functions $f_1$ and $f_2$ from the conjugate classes of $(\psi, \alpha)$-holomorphy such that

$$f = f_1 + f_2.$$

Analogously for $\lambda \in \mathfrak{G}$ there exist three functions $g_1, g_2, g_3$ with the properties: $g_1$ is $(\psi, 2\alpha_0)$-holomorphic, $g_2$ is from the conjugate class, $g_3$ is harmonic (!), and such that

$$f = g_1\alpha + g_2\alpha + g_3\vec{\alpha}.$$

It is easy to write down the formulas which express how $f_1, f_2$ as well as $g_1, g_2, g_3$ are determined via $f$.

What is most paradoxical here is that if $f$ is $\mathbb{C}$-valued or even $\mathbb{R}$-valued then nevertheless both of the above representations for $f$ are valid. But of course all functions arising ($f_1, f_2$ and $g_1, g_2, g_3$) are quaternion-valued!

Let us compare this with the harmonic case. Real-valued harmonic functions in the plane are closely related to holomorphic functions which are necessarily complex-valued, that is, we are forced to enlarge appropriately the number system. Complex-valued harmonic functions does not require, evidently, any more enlarging. But considering then metaharmonic functions, no matter whether they are real or complex, we arrive at the necessity of employing a larger, less conventional number system, that of the quaternions. Maybe this explains the absence of an analysis, related to metaharmonic functions, until very recently.

**A.4.7** If we take different components in $\bar{\partial}_{\psi,\alpha}[f] = 0$ we get the corresponding Cauchy–Riemann conditions

$$\begin{cases} \alpha_0 f_0 - \sum_{j=1}^{2}(\sum_{k=1}^{2} \psi_j^k \cdot \partial_k f_j + \alpha_j f_j) = 0, \\ \alpha_0 f_1 + \alpha_1 f_0 - \alpha_2 f_3 + \sum_{k=1}^{2}(\psi_1^k \partial_k f_0 - \psi_3^k \cdot \alpha_k f_2 + \psi_k^2 \cdot \partial_k f_3) = 0, \\ \alpha_2 f_0 + \alpha_0 f_2 - \alpha_3 f_1 + \sum_{k=1}^{2}(\psi_2^k \cdot \partial_k f_0 + \psi_3^k \cdot \partial_k f_1 - \psi_1^k \partial_k f_3) = 0, \\ \alpha_3 f_0 + \alpha_0 f_3 + \alpha_2 f_1 + \sum_{k=1}^{2}(\psi_3^k \cdot \partial_k f_0 - \psi_2^k \cdot \partial_k f_1 + \psi_1^k \partial_k f_2) = 0. \end{cases}$$

For some applications it is more convenient to express the conditions of $\alpha$-holomorphy in vectorial terms. It is natural to consider now two-dimensional vectors of the form $\vec{g} = g_1 i_1 + g_2 i_2$. So let $\psi = \psi_{st} = (i_1, i_2)$, $f = \sum_{k=0}^{3} f_k i_k = (f_0 + f_3 i_3) + (f_1 i_1 + f_2 i_2) = (f_1 i_1 + f_2 i_2) + (-f_0 i_2 + f_3 i_1) i_2 =: \vec{F}_1 + \vec{F}_2 \cdot i_2$; $\alpha = \vec{A}_1 + \vec{A}_2 \cdot i_2$.

Vector operations take the form:

$$< \vec{g}, \vec{h} >= g_1 h_1 + g_2 h_2 \; ; \qquad [\vec{g}, \vec{h}] = i_1 i_2 (g_1 h_2 - g_2 h_1);$$

$$\mathrm{div}\vec{g} = \partial_1 g_1 + \partial_2 g_2, \qquad \mathrm{rot}\vec{g} = i_1 i_2 (\partial_1 g_2 - \partial_2 g_1), \qquad \mathrm{grad} g_0 = i_1 \partial_1 g_0 + i_2 \partial_2 g_0.$$

Vector operations and quaternionic multiplication are related as follows. If $f = \vec{F}_1 + \vec{F}_2 \cdot i_2$, $g = \vec{G}_1 + \vec{G}_2 i_2$ then

$$
\begin{aligned}
f\cdot g &= -<\vec{F}_1,\vec{G}_1> +[\vec{F}_1,\vec{G}_1]+<\vec{F}_2,\vec{G}_2^*> -[\vec{F}_2,\vec{G}_2^*]+\\
&+ (-<\vec{F}_1,\vec{G}_2> +[\vec{F}_1,\vec{G}_2]-<\vec{F}_2,\vec{G}_1^*> +[\vec{F}_2,\vec{G}_1^*])i_2 =\\
&= (([\vec{F}_1,\vec{G}_2]^\wedge + [\vec{F}_2,\vec{G}_1^*]^\wedge) - (<\vec{F}_1,\vec{G}_2> + <\vec{F}_2,\vec{G}_1^*>)i_2)+\\
&+ (-([\vec{F}_1,\vec{G}_1]^\wedge - [\vec{F}_2,\vec{G}_2^*]^\wedge) + (<\vec{F}_1,\vec{G}_1> - <\vec{F}_2,\vec{G}_2^*>)i_2)i_2,
\end{aligned}
$$

where

$$[\vec{g},\vec{h}]^\wedge := [\vec{g},\vec{h}]\cdot i_2 = -i_1(g_1h_2 - g_2h_1)$$

is a "rotated" vector product and where

$$\vec{g}^* := -g_1i_1 + g_2i_2$$

if $\vec{g} = g_1i_1 + g_2i_2$.

Then after a straightforward calculation we get:

$$
\begin{aligned}
\partial_\alpha[f] = \partial_\alpha[\vec{F}_1 + \vec{F}_2 i_2] &= (i_1\partial_1 + i_2\partial_2)(\vec{F}_1 + \vec{F}_2 i_2) + (\vec{F}_1 + \vec{F}_2 i_2)(\vec{A}_1 + \vec{A}_2 i_2) =\\
&= -\mathrm{div}\vec{F}_1 + \mathrm{rot}\vec{F}_1 + (-\mathrm{div}\vec{F}_2 + \mathrm{rot}\vec{F}_2)i_2+\\
&- <\vec{F}_1,\vec{A}_1> + <\vec{F}_2,\vec{A}_2^*> -[[\vec{F}_2,\vec{A}_2^*] - [\vec{F}_1,\vec{A}_1]]+\\
&+ [-<\vec{F}_1,\vec{A}_2> - <\vec{F}_2,\vec{A}_1^*> +[\vec{F}_1,\vec{A}_2] + [\vec{F}_2,\vec{A}_1^*]]i_2 =\\
&= ((\mathrm{rot}^\wedge\vec{F}_2 + [\vec{F}_1,\vec{A}_1]^\wedge + [\vec{F}_2,\vec{A}_1^*]^\wedge)-\\
&- (\mathrm{div}\vec{F}_2 + <\vec{F}_1,\vec{A}_2> + <\vec{F}_2,\vec{A}_1^*>)i_2)+\\
&+ ((-\mathrm{rot}^\wedge\vec{F}_1 + [\vec{F}_2,\vec{A}_2^*]^\wedge - [\vec{F}_1,\vec{A}_1]^\wedge)+\\
&+ (\mathrm{div}\vec{F}_1 + <\vec{F}_1,\vec{A}_1> - <\vec{F}_2,\vec{A}_2^*>)i_2)i_2,
\end{aligned}
$$

where

$$\mathrm{rot}^\wedge\vec{g} := \mathrm{rot}\vec{g}\cdot i_2 = -i_1(\partial_1 g_2 - \partial_2 g_1)$$

is the "rotated" vector $\mathrm{rot}\vec{g}$.

This implies that $f$ is $\alpha-$ holomorphic if and only if

$$\begin{cases} \text{rot}\vec{F}_2 + [\vec{F}_1, \vec{A}_1] + [\vec{F}_2, \vec{A}_1^*] = 0, \\ \text{div}\vec{F}_2 + < \vec{F}_1, \vec{A}_2 > + < \vec{F}_2, \vec{A}_1^* >= 0, \\ [\vec{F}_2, \vec{A}_2^*] - [\vec{F}_1, \vec{A}_1] - \text{rot}\vec{F}_1 = 0, \\ \text{div}\vec{F}_1 + < \vec{F}_1, \vec{A}_1 > - < \vec{F}_2, \vec{A}_2^* >= 0. \end{cases}$$

In particular if $\alpha = 0$; i.e., $\vec{A}_1 = \vec{A}_2 = 0$, then we arrive at

$$\text{rot}\vec{F}_1 = 0, \qquad \text{div}\vec{F}_1 = 0,$$

$$\text{rot}\vec{F}_2 = 0, \qquad \text{div}\vec{F}_2 = 0,$$

which is equivalent, for $F_1$, $F_2$ being $\mathbb{R}$-valued vectors, to the usual Cauchy–Riemann conditions of holomorphy of the functions $F_1$ and $F_2$.

# Bibliography


[1] M. Abramowitz, I. A. Stegun, *Handbook of mathematical functions; with formulas, graphs, and mathematical tables*. N.Y.: Dover Publ., 1965, 1045 pp.

[2] A. I. Akhijezer, I. A. Akhijezer, *Electromagnetism and electromagnetic waves*. Moscow: Visshaya shkola, 1985 (in Russian).

[3] M. A. Alexidze, *Fundamental functions in approximate solutions of boundary value problems*. Moscow: Nauka, 1991, 352 pp. (in Russian).

[4] B. D. Annin, Yu. M. Grigor'ev, V. V. Naumov, *Solution of spatial static problems of elasticity theory by methods of quaternionic function theory*. 9 All-Union conference on approximate methods for solving of problems of elasticity theory, Novosibirsk, 1986, 35–42 (in Russian).

[5] V. E. Balabaev, *A certain class of multidimensional elliptic systems of first order*. Diferentsial'nye Uravneniya, 1992, v. 28, # 4, 628–637 (in Russian); Engl. transl.: Differential equations, 1992, v. 28, # 4, 510–517.

[6] V. E. Balabaev, *Boundary value problems for canonical systems of first order*. Differentsial'nye Uravneniya, 1993, v. 29, # 8, 1358–1369 (in Russian); Engl. transl.: Differential Equations,

[7] V. B. Berestetskij, E. M. Lifshits, L. P. Pitaevskij, *Quantum electrodynamics*. Moscow: Nauka, 1989, 728 pp. (in Russian). 1993, v. 29, # 8, 1177–1187.

[8] A. V. Berezin, Yu. A. Kurochkin, E. A. Tolkachev, *Quaternions in relativistic physics*. Minsk: Nauka y Tekhnika, 1989 (in Russian).

[9] A. V. Berezin, E. A. Tolkachev, F. I. Fedorov, *Lorentz transformations and equations for spinor quaternions*. Doklady Akademii Nauk BSSR, 1980, v. 24, # 4, 308–310 (in Russian).

[10] S. Bergman, *Integral operators in the theory of linear partial differential equations*. N.Y.: Springer–Verlag, 1969.

[11] S. Bernstein, *Fundamental solutions for Dirac–type operators in Clifford analysis.* Abstracts of the Banach Center Symposium "Generalizations of complex analysis and their applications in physics". Part 1, 1–2, Warsaw, 1994.

[12] S. Bernstein, *The left–linear Riemann problem in Clifford analysis as an integral equation.* Preprint 95–04, Fakultät für Mathematik und Informatik, Technische Universität Bergakademie Freiberg, 25 pp.

[13] R. K. Bhaduri, *Models of the nucleon: from quarks to soliton.* Addison–Wesley Publ. Co., Inc., 1988.

[14] A. V. Bitsadze, *Boundary value problems for second–order elliptic equations.* Amsterdam: North–Holland and N.Y.: Interscience, 1968.

[15] A. V. Bitsadze, *Two–dimensional analogs of the Hardy and Hilbert inversion formulas.* Doklady Akademii Nauk, Russia, 1993, v. 333, # 6, 696–698.

[16] N. N. Bogoliubov, D. V. Shirkov, *Introduction to the theory of quantized fields.* Moscow: Nauka, 1984, 600 pp. (in Russian).

[17] B. V. Bojarskij, *Some boundary value problems for elliptic type equations in the plane.* Ph.D. Dissertation, Moscow State University, 1955.

[18] F. Brackx, R. Delanghe, F. Sommen, *Clifford analysis.* London: Pitman Res. Notes in Math., v. 76, 1982, 308 pp.

[19] F. Brackx, N. Van Acker, *Boundary value theory for eigenfunctions of the Dirac operator.* Bull. Soc. Math. Belg., 1993, v.45, # 2, Ser. B, 113-123.

[20] L. Bragg, J. Dettman, *Function theories for the Yukawa and Helmholtz equation.* Rocky Mountain J. Math., 1995, v. 25, # 3, 887–917.

[21] Yu. A. Brichkov, A. P. Prudnikov, *Integral transforms of distributions.* Moscow: Nauka, 1977, 288 pp. (in Russian).

[22] G. Casanova, *L'algebre vectorielle.* Presses Universitaires de France, 1976.

[23] A. Chodos, R. L. Jaffe, K. Johnson, C. B. Thorn, *Baryon structure in the bag theory.* Phys. Rev. D, 1974, v. 10, 2599–2604.

[24] A. Chodos, R. L. Jaffe, C. B. Thorn, V. Weisskopf, *New extended model of hadrons.* Phys. Rev. D, 1974, v. 9, 3471–3495.

[25] S. Chumakov, K. B. Wolf, *Supersymmetry in Helmholtz optics.* Phys. Lett. A, 1994, v. 193, 51–53.

[26] F. Close, *An introduction to quarks and partons.* N.Y.: Academic Press, 1979, 495 pp.

[27] D. Colton, R. Kress, *Integral equations methods in scattering theory.* N. Y.: John Wiley and Sons, 1983.

[28] D. Colton, R. Kress, *Inverse acoustic and electromagnetic scattering theory.* Berlin: Springer, 1992.

[29] R. Delanghe, F. Sommen, V. Soucek, *Clifford algebra and spinor–valued functions.* Amsterdam: Kluwer Acad. Publ., 1992, 485 pp.

[30] A. A. Dezin, *Invariant differential operators and boundary value problems.* Trudi Matematicheskogo instituta im. V.A.Steklova, Moscow, 1962, v. 68 (in Russian).

[31] G. Dixon, *Division algebras: octonions, quaternions, complex numbers, and the algebraic design of physics.* Dordrecht: Kluwer, 1994.

[32] V. I. Dmitriev, E. V. Zakharov, *Integral equations in boundary value problems of electrodynamics.* Moscow: Moscow State University Publ., 1987, 167 pp. (in Russian).

[33] V. G. Drinfel'd, V. V. Sokolov, *Equations related to the Korteweg–de Vries equation.* Dokl. Akad. Nauk SSSR, 1985, v.284, 29–33 (in Russian); Engl. transl. Soviet Math. Dokl., 1985, v. 32, 361–365.

[34] A. D. Dzhuraev, *Singular integral equation method.* Moscow: Nauka, 1987 (in Russian); Engl. transl. Longman Sci. Tech., Harlow and Wiley, N.Y., 1992.

[35] E. A. Ermolaev, *Some applications of quaternions of rank r in the theory of relativistic wave equations.* Preprint # 435, Institute of Physics, Academy of Sciences of BSSR, 1986, 36 pp. (in Russian).

[36] B. V. Fedosov, *Analytical formulas for index of elliptic operators.* Trudi Moskovskogo matematishcheskogo obshestva, 1974, v. 20, 159–241 (in Russian).

[37] R. Füter, *Analytische Funktionen einer Quaternionen variablen.* Comment Math. Helv., 1932, v. 4, 9–20.

[38] F. D. Gakhov, *Boundary value problems.* Moscow: Nauka, 1977, 640 pp (3d Edition); Engl. transl. of the first Edition, Oxford: Pergamon, 1966.

[39] D. V. Galtsov, Yu. V. Grats, V. Ch. Zhukovskij, *Classical fields.* Moscow: University Press, 1991, 150 pp. (in Russian).

[40] J. E. Gilbert, M. A. M. Murray, *Clifford algebras and Dirac operators in harmonic analysis.* Cambridge Univ. Press, Cambridge studies in Adv. Math., v. 26, 1991, 334 pp.

[41] P. Gonzalez-Casanova, K. B. Wolf, *Interpolation for solutions of the Helmholtz equation.* Numerical Methods for P.D.E., 1995, v. 11, 77–91.

[42] Yu. M. Grigor'ev, *Some solutions of spatial static Lame equations.* Dinamika sploshnoj sredi, Novosibirsk, 1984, v. 67, 29–36 (in Russian).

[43] Yu. M. Grigor'ev, V. V. Naumov, *Approximation theorems for the Moisil-Theodoresco system.* Sibirskii Matematicheskii Zhurnal, 1984, v. 25, # 5, 9-19 (in Russian); Engl. transl.: Siberian Mathematical Journal, 1984, v. 25, # 5, 693-701.

[44] I. S. Gudovich, *On holomorphic vector-functions depending on an arbitrary number of real variables.* In:"Primenenije novikh metodov analiza v teorii kraevikh zadach". Voronezh, Voronezh State University, 1990, 5–11 (in Russian).

[45] K. Gürlebeck, *Hypercomplex factorization of the Helmholtz equation.* Zeitschrift für Analysis und ihre Anwendungen, 1986, Bd. 5(2), 125–131.

[46] K. Gürlebeck, W. Sprößig, *Quaternionic analysis and elliptic boundary value problems.* Berlin: Akademie-Verlag, 1989.

[47] F. Gürsey, H. C. Tze, *Complex and quaternionic analyticity in chiral and gauge theories. Part 1.* Ann. Phys., 1980, v. 128, 29–130.

[48] R. F. Harrington, *Time-harmonic electromagnetic fields.* N.Y. McGraw-Hill, 1961.

[49] W. A. Hurwitz, *Über die Komposition der quadratischen Formen.* Mathematische Annalen, 1922, v. 88, # 1/2, 1–25.

[50] A. S. Iljinskij, V. V. Kravtsov, A. G. Sveshnikov, *Mathematical models of electrodynamics.* Moscow: Visshaya Shkola, 1991 (in Russian).

[51] K. Imaeda, *A new formulation of classical electrodynamics.* Nuovo Cimento, 1976, v. 32 B, # 1, 138–162.

[52] B. Jancewicz, *Multivectors and Clifford algebra in electrodynamics.* Singapore: World Scientific, 1988.

[53] B. Jawerth, M. Mitrea, *Acoustic scattering Galerkin estimates and Clifford algebras.* Clifford Algebras in Analysis and related topics, J. Ryan (ed.), N.Y.: CRC Press, 1995, 199–216.

[54] D. S. Jones, *Acoustic and electromagnetic waves.* N. Y.: Clarendon Press. Oxford, 1986.

[55] I. L. Kantor, A. S. Solodovnikov, *Hypercomplex numbers*. Springer, 1989.

[56] V. V. Kisil, *Connection between different function theories in Clifford analysis*. Advances in Applied Clifford Algebras, 1995, v. 5, # 1, 63–74.

[57] H. König, *An explicit formula for fundamental solutions of linear partial differential equations with constant coefficients*. Proc. Am. Math. Soc., 1994, v. 120, # 4, 1315–1318.

[58] V. G. Kravchenko, V. V. Kravchenko,*On some nonlinear equations generated by Fueter type operators*. Zeitschrift für Analysis und ihre Anwendungen, 1994, v. 13, # 4, 599–602.

[59] V. G. Kravchenko, M. V. Shapiro, *Hypercomplex factorization of the multidimensional Helmholtz operator and some of its applications*. Doklady rasshirennyh zasedaniy seminara instituta prikl. mat. imeni I. N. Vekua, Tbilisi, 1990, v. 5, # 1, 106–109 (in Russian).

[60] V. V. Kravchenko, *On the relation between holomorphic biquaternionic functions and time-harmonic electromagnetic fields*. Deposited in UKRINTEI 29.12.1992 under Nr. 2073-Uk-92 (in Russian).

[61] V. V. Kravchenko, *On a hypercomplex factorization of some equations of mathematical physics*. Preprint # 1, Dep. of Math., Instituto Superior Técnico, Lisbon, Portugal, 1993, 8 pp.

[62] V. V. Kravchenko, *Generalized holomorphic vectors and $\alpha$-hyperholomorphic function theory*. Journal of Natural Geometry, 1994, v. 6, # 2, 125–132.

[63] V. V. Kravchenko, *Integral representations of biquaternionic hyperholomorphic functions and their applications*. Ph.D. Dissertation, Odessa State University, 1993, 95 pp. (in Russian).

[64] V. V. Kravchenko, *On a biquaternionic bag model.* Zeitschrift für Analysis und ihre Anwendungen, 1995, v.14, # 1, 3–14.

[65] V. V. Kravchenko, *Quaternion-valued integral representations for time-harmonic electromagnetic and spinor fields*. Doklady Akademii Nauk, Russia, 1995, v. 341, # 5, 603–605; Engl. translation in Russian Academy of Sciences. Doklady.

[66] V. V. Kravchenko, *On a decomposition into a direct sum of the Laplace operator kernel by means of biquaternionic zero divisors*. Differentsialnye Uravneniya,1995, v. 31, # 3, 498–501 (in Russian); Engl. translation: Differential Equations.

[67] V. V. Kravchenko, E. Ramirez de Arellano, M. V. Shapiro, *On integral representations and boundary properties of spinor fields.* Preprint, Dep. of Mathematics, CINVESTAV del IPN, Mexico City, Mexico, 1995, 20 pp. Mathematical Methods in the Applied Sciences, 1996, v. 19.

[68] V. V. Kravchenko, G. Santana, *On singular integral operators associated with the vector Helmholtz equation.* Journal of Natural Geometry, 1996, v. 10, # 2, 119–136.

[69] V. V. Kravchenko. M. V. Shapiro, *Helmholtz operator with a quaternionic wave number and associated function theory. II. Integral representations.* Acta Applicandae Mathematicae, 1993, v. 32, # 3, 243–265.

[70] V. V. Kravchenko, M. V. Shapiro, *On the generalized system of Cauchy–Riemann equations with a quaternion parameter.* Doklady Akademii Nauk, Russia, 1993, v. 329, # 5, 547–549 (in Russian), English transl.: Russian Acad. Sci. Dokl. Math. 1993, v 47, # 2, 315–319.

[71] V. V. Kravchenko, M. V. Shapiro, *Helmholtz operator with a quaternionic wave number and associated function theory.* Deformations of Mathematical Structures, II. Kluwer Academic Publishers (ed.: J. Lawrynowicz), 1994, 101–128.

[72] V. V. Kravchenko, M. V. Shapiro, *Quaternionic time–harmonic Maxwell operator.* Journal of Physics A: Math. Gen., 1995, v. 28, 5017–5031.

[73] N. Ya. Krupnik, *Banach algebras with symbol and singular integral operators.* Kishinev: Shtiinca, 1984, 138 pp. (in Russian).

[74] V. D. Kupradze, T. G. Gegelia, M. O. Bashaleishvili, T. V. Burchuladze, *Three-dimensional problems of mathematical theory of elasticity and thermoelasticity.* Moscow: Nauka, 1976 (in Russian).

[75] V. N. Kutrunov, *Quaternionic method for regularizing the integral equations of elasticity theory.* Prikl. Matematika y Mekhanika, 1992, v. 56, # 5, 864–868 (in Russian).

[76] A. M. Kytmanov, *Bochner–Martinelli integral and its applications.* Moscow: Nauka, 1992, 244 pp. (in Russian).

[77] Huang Liede, *The existence and uniqueness theorem of the linear and nonlinear Riemann–Hilbert problems for the generalized holomorphic vector of the second kind.* Acta Mat. Sci. Engl. Ed., 1990, v. 10, # 2, 185–199.

[78] G. S. Litvinchuk, *Boundary value problems and singular integral equations with shift.* Moscow: Nauka, 1977, 448 pp. (in Russian).

[79] H. R. Malonek, *Hypercomplex differentiability and its applications.* Clifford algebras and applications in mathematical physics, F. Brackx et al., eds., Kluwer Acad. Publ., Netherlands, 1993, 141–150.

[80] A. V. Mikhailov, A. B. Shabat, R. I. Yanilov, *Extension of the module of invertible transformations.* Comm. Math. Phys., 1988, v. 115, 1–19.

[81] S. G. Mikhlin, S. Prößdorf, *Singular integral operators.* Springer–Verlag, 1986, 528 pp.

[82] I. M. Mitelman, M. V. Shapiro, *Formulae of changing of integration order and of inversion for some multidimensional singular integrals and hypercomplex analysis.* Journal of Natural Geometry, 1994, v. 5, 11-27.

[83] I. M. Mitelman, M. V. Shapiro, *Differentiation of the Martinelli–Bochner integrals and the notion of hyperderivability.* Mathematische Nachrichten, 1995, v. 172, 211–238.

[84] M. Mitrea, *Boundary value problems and Hardy spaces associated to the Helmholtz equation in Lipschitz domains.* Res. Report 1992:02, Dep. of Math., University of South Carolina.

[85] M. Mitrea, *Clifford wavelets, singular integrals, and Hardy spaces.* N.Y.: Springer Verlag, Lecture Notes, v. 1575, 1994.

[86] S. Mizohata, *Theory of partial differential equations.* Moscow: Mir, 1977, 504 pp. (Russian edition).

[87] G. Moisil, *Sur les quaternions monogenes.* Bull. Sci. Math. Paris, 1931, v. 55, # 2, 169–194.

[88] G. Moisil, N. Theodoresco, *Functions holomorphes dans l'espace.* Mathematica (Cluj), 1931, v. 5, 142–159.

[89] N. Morita, N. Kumagai, J. R. Mautz, *Integral equation methods for electromagnetics.* Norwood: Artech House, 1990.

[90] V. M. Mostepanenko, N. N. Trunov, *Casimir's effect and its applications.* Moscow: Energoatomizdat, 1990, 216 pp. (in Russian).

[91] C. Müller, *Grundprobleme der mathematischen Theorie elektromagnetischer Schwingungen.* Berlin: Springer–Verlag, 1957.

[92] V. V. Nikolski, *Electromagnetics and propagation of radiowaves.* Moscow: Nauka, 1978, 544 pp. (in Russian).

[93] K. Nôno, *On the quaternion linearization of Laplacian* Δ . Bull. Fukuoka Univ. Educ. Nat. Sci., 1985, v. 35, 5–10.

[94] E. I. Obolashvili, *Three-dimensional generalized holomorphic vectors.* Differentsialnye. Uravneniya, 1975, v. 11, # 1, 108–115 (in Russian), English transl.: Differential Equations, 1975, v. 11, # 1, 82–87.

[95] P. J. Olver, *Applications of Lie groups to differential equations.* N.Y.: Springer–Verlag, 1986, 523 pp.

[96] W. Pauli, *Theory of relativity.* Moscow: Nauka, 1991 (Russian edition).

[97] W. M. Pezzaglia Jr., *Multivector solutions to harmonic systems.* Clifford Algebras and Their Applications in Mathematical Physics (eds.: J.S.R.Chisholm and A.K.Common), 1986, p. 445–454.

[98] L. S. Pontriagin, *Generalizations of numbers.* Moscow, 1986 (in Russian).

[99] S. Prössdorf, J. Saranen, *A fully discrete approximation method for the exterior Neumann problem of the Helmholtz equation.* Zeitschrift Anal. Anwend., 1994, v. 13, # 4, 683–695.

[100] J. Ryan, *Cauchy–Green type formulae in Clifford analysis.* Transactions of the AMS, 1995, v. 347, 1331–1341.

[101] J. Ryan (editor), *Clifford algebras in analysis and related topics.* N.Y.: CRC Press, 1995.

[102] J. Ryan, *Intrinsic Dirac operators in* $\mathbb{C}^n$ . Advances in Mathematics, 1996, v. 118, 99–133.

[103] S. G. Samko, A. A. Kilbas, O. I. Marichev, *Integrals and derivatives of a fractional order and some of their applications.* Minsk: Nauka y Tekhnika, 1987, 688 pp. (in Russian).

[104] I. R. Shafarevich, *Basic notions of algebra.* Sovremennye problemi matematiki. Fundamentalnye napravleniya, v. 11, Moscow: VINITI, 1985 (in Russian); Engl. transl. Encyclopedia of Mathematical Sciences, v. 11 Algebra I, Berlin: Springer–Verlag, 1990.

[105] M. V. Shapiro, *On the properties of a class of singular integral equations connected with the three–dimensional Helmholtz equation.* Abstracts of lectures at the 14th School on operator theory in functional spaces, Novgorod, USSR, 1989, p. 88 (in Russian).

[106] M. V. Shapiro, *On an analogue of the Toeplitz operators.* Linear operators in functional spaces, Grozni, USSR, 1989, 178–179 (in Russian).

[107] M. V. Shapiro, *On analogies of the Riemann boundary value problem for a class of hyperholomorphic functions.* In: "Integral equations and boundary value problems", World Scientific, 1991, 184–188.

[108] M. V. Shapiro, *Some remarks on generalizations of one-dimensional complex analysis: hypercomplex approach.* Preprint # 131, June 1993, Dep. of Math., CINVESTAV del I.P.N., Mexico-city, Mexico, 23 pp. Also to appear in: Proceedings of the second workshop on functional-analytic methods in complex analysis and applications to partial differential equations, Trieste, Jan. 1993.

[109] M. V. Shapiro, L. M. Tovar, *Two-dimensional Helmholtz operator and its hyperholomorphic solutions.* Journal of Natural Geometry, 1997 (to appear).

[110] M. V. Shapiro, L. M. Tovar, *On a class of integral representations related to the two-dimensional Helmholtz operator.* To appear.

[111] M. V. Shapiro, N. L. Vasilevski, *Quaternionic $\psi$-hyperholomorphic functions, singular integral operators with quaternionic Cauchy kernel and analogues of the Riemann boundary value problem.* Preprint # 102, Dep. of Math., CINVESTAV del IPN, Mexico City, Mexico, 1992, 75 pp.

[112] M. V. Shapiro, N. L. Vasilevski, *On the Bergman kernel function in hyperholomorphic analysis.* Acta Applicandae Mathematicae (to appear).

[113] M. V. Shapiro, N. L. Vasilevski, *Quaternionic $\psi$–hyperholomorphic functions, singular integral operators and boundary value problems. I. $\psi$–hyperholomorphic function theory.* Complex Variables. Theory and Applications, 1995, v. 27, 17–46.

[114] M. V. Shapiro, N. L. Vasilevski, *Quaternionic $\psi$–hyperholomorphic functions, singular integral operators and boundary value problems. II. Algebras of singular integral operators and Riemann type boundary value problems.* Complex Variables. Theory and Applications, 1995, v. 27, 67–96.

[115] V. I. Shevchenko, *On the Hilbert problem for a holomorphic vector in a multidimensional space.* Differentsialnye y integralnye uravneniya, Tbilisi: University Publ., 1979, 279–291 (in Russian).

[116] M. S. Shneerson, *On Moisil monogenic functions.* Mat. Sbornik, 1958, v. 44, # 1, 113–122 (in Russian).

[117] V. V. Sokolov, *On the symmetries of evolution equations.* Uspekhi Mat. Nauk, 1988, v. 43, 5, 133–163 (in Russian); Engl. transl.: Russian Math. Surveys, 1988, v. 43, 5, 165–204.

[118] M. Z. Solomjak, *On linear elliptic systems of first order.* Doklady Akademii Nauk SSSR, 1963, v. 150, #1, 48–51 (in Russian).

[119] V. Soucek, *Complex–quaternionic analysis applied to spin 1/2 massless fields.* Complex Variables, Theory and Applications, 1983, v. 1, # 4, 327–346.

[120] W. Sprößig, *On the treatment of non–linear boundary value problems of a disturbed Dirac equation by hypercomplex methods.* Complex Variables, 1993, v. 23, 123–130.

[121] E. Stein, G. Weiss, *Introduction to Fourier analysis on Euclidean spaces.* Princeton: Princeton University Press, 1971, 307 pp.

[122] S. I. Svinolupov, V. V. Sokolov, *Factorization of evolution equations.* Uspekhi Mat. Nauk, 1992, v. 47, 3, 115–146 (in Russian); Engl. transl.: Russian Math. Surveys, 1992, v. 47, 3, 127–162.

[123] N. N. Tarkhanov, *A remark on the Moisil–Teodorescu system.* Sibirskii Matematicheskii Zhurnal, 1987, v. 28, # 3, 208–213 (in Russian); Engl. transl.: Siberian Mathematical Journal, 1987, v. 28, # 3, 518–522.

[124] A. W. Thomas, *Chiral symmetry and the bag model. A new starting point for nuclear physics.* Adv. Nucl. Phys., 1984, v. 13, 1–137.

[125] A. M. Tsalik, *Quaternionic functions, their properties and some applications to the problems of mechanics.* Doklady Akademii Nauk UkrSSR, Series A, 1986, # 12, 21–24 (in Russian).

[126] N. L. Vasilevski, M. V. Shapiro, *On an analogue of monogenicity in the sense of Moisil–Theodoresco and some applications in the theory of boundary value problems.* Reports of Enlarged Session of Seminars of the I.N.Vekua Institute of Appl. Math., Tbilisi, 1985, v. 1, 63–66 (in Russian).

[127] N. L. Vasilevski, M. V. Shapiro, *Quaternionic $\psi$–monogenic functions, singular operators with the quaternionic Cauchy kernel, and analogs of the Riemann problem.* Deposited in UkrNINTI 06.02.1987, # 620-Uk-87, 68 pp. (in Russian).

[128] A. M. Vinogradov, I. S. Krasil'shik, V. V. Lychagin, *Introduction to geometry of non–linear differential equations.* Moscow: Nauka, 1986 (in Russian).

[129] V. S. Vinogradov, *Spinor systems.* Differentsial'nye Uravneniya, 1991, v. 27, #1, 22–29 (in Russian); Engl. transl.: Differential equations, 1991, v. 27, # 1, 17–22.

[130] V. S. Vladimirov, *Equations of mathematical physics.* Moscow: Nauka, 1984 (in Russian); Engl. transl. of the first edition: N.Y.: Marcel Dekker, 1971.

[131] Xu Zhenyuan, *A function theory for the operator D-$\lambda$*. Complex Variables, Theory and Appl. 1991, v. 16, 27–42.

[132] A. Yanushauskas, *Some generalizations of holomorphic vectors*. Differentsial'nye Uravneniya, 1982, v. 18, # 4, 699–705 (in Russian); Engl. transl.: Differential Equations, 1982, v. 18, # 4, 517–523.

[133] A. Yanushauskas, *On some mappings realized by harmonic functions*. Lit. Math. Sbornik, 1989, v. 29, 819–825 (in Russian).

[134] A. Yanushauskas, *Multidimensional elliptic systems with varying coefficients*. Vilnius: Mokslas, 1990, 179 pp. (in Russian).

[135] M. S. Zhdanov, *Analogues of the Cauchy–type integral in the theory of geophysical fields*. Moscow: Nauka, 1984 (in Russian).

[136] M. S. Zhdanov, *Integral transforms in geophysics*. Heidelberg: Springer-Verlag, 1988.

[137] M. S. Zhdanov, N. L. Vasilevski, M. V. Shapiro, *The space analogues of the Cauchy type integral and the theory of quaternions*. Academy of Sciences of the USSR Institute of Earth Magnetism, Ionosphere and Radio–Waves Propagation. Preprint # 48 (737). Moscow, 1987, 23 pp. (in Russian).

# Index

T - #0224 - 071024 - C0 - 241/165/14 - PB - 9780582297418 - Gloss Lamination